It has been shown that we learn by doing. Perhaps engineering students especially will better understand the principles of heat transfer and thermodynamics by conducting experiments and seeing results.

This book presents a collection of experiments in heat transfer and thermodynamics contributed by leading engineering educators. The experiments have been tested, evaluated, and proved to be successful for classroom use. They are fun and challenging.

Each experiment follows a similar step-by-step format, which includes the objective of the experiment, apparatus needed, procedure, suggested headings, and references. The experiments use apparatus that is easily built or obtained.

Among the topics covered are heat conduction, convection, boiling, mixing, diffusion, radiation, heat pipes and exchangers, and thermodynamics. Appendixes include lists of short experiments and demonstrations that have appeared in the literature, along with lists of available films and audiovisual materials and where to get them.

Designed to serve as a companion to standard heat transfer and thermodynamics texts, this book will be a useful and appealing resource for engineering students.

Experiments in heat transfer and thermodynamics

Experiments
in heat transfer
and thermodynamics

Edited by

Robert A. Granger

Professor of Mechanical Engineering
United States Naval Academy

CAMBRIDGE UNIVERSITY PRESS
Cambridge, New York, Melbourne, Madrid, Cape Town, Singapore,
São Paulo, Delhi, Dubai, Tokyo, Mexico City

Cambridge University Press
The Edinburgh Building, Cambridge CB2 8RU, UK

Published in the United States of America by
Cambridge University Press, New York

www.cambridge.org
Information on this title: www.cambridge.org/9780521449250

First published 1994

A catalogue record for this publication is available from the British Library

Library of Congress Cataloguing in Publication Data

Experiments in heat transfer and thermodynamics / edited by Robert A. Granger.
p.   cm.
ISBN 0-521-45115-9 (hardback). – ISBN 0-521-44925-1 (pbk.)
1. Heat – Transmission – Experiments.   2. Thermodynamics–Experiments.
I. Granger, Robert Alan.
TJ260.E97   1994
621.402′2′078 – dc20   93-29702
                                                                          CIP

ISBN 978-0-521-45115-4 Hardback
ISBN 978-0-521-44925-0 Paperback

Man should have heroes. This book is dedicated to mine.

To **Stephen W. Hawking** of Cambridge University, who on contemplating the origin of the universe stated, "Entropy and disorder increase with time because we measure time in the direction in which disorder increases."*

To **John H. Nickerson** of Stamford, CT, who makes order out of disorder.

* "Quantum cosmology," *300 Years of Gravitation*, eds. S. W. Hawking and W. Israel, Cambridge University Press, Cambridge, 1987, p. 648.

Contents

Preface

This book represents a collection of the favorite experiments in heat transfer and thermodynamics of some of the world's eminent scholars in the field. It was my intent to include experiments that they had used in their lectures which had a proven track record in regard to students' reaction and understanding. Experiments that were too complicated, even bordering on research, or too esoteric were rejected.

Some of the experiments are well known, some of them are relatively new; and for all of them I have tried to devise the presentation that appears to be best from a consistent point of view. The book has been prepared as a connected account; each experiment is intended to be used as a whole rather than in parts. The readers were envisioned as both engineering and physics students; however, the appeal will probably lie with engineering students from all disciplines.

A brief word is necessary about the selection of topics to be found in this book as well as the order in which they were placed. The experiments selected were varied, representative, new, old, fascinating, and challenging. I selected those I thought my students would enjoy as well as students in Japan, Germany, Brazil, and other countries with different backgrounds and methodologies.

As to the order of the material, the book is partitioned into experiments in heat transfer and then experiments in thermodynamics. Most schools do not have a one-to-two-hour laboratory solely in thermodynamics. Thus there are only seven experiments in thermodynamics, but what wonderful experiments they are! Appendix 1 presents a list of demonstrations or short experiments for those who desire the more classic demonstrations in thermodynamics.

The main emphasis of the experiments in this book is on heat transfer. In Part I, there are 25 experiments distributed among the fields of boiling, condensation, conduction, convection (both free and forced), radiation, heat pipes, exchangers, mixing, dispersion, and diffusion. As in the thermodynamics section there is a list of short experiments and demonstrations (Appendix 2) in heat transfer that I have found can spice up a lengthy theoretical derivation.

One may also find videos and films a refreshing substitute for the laboratory. Appendix 3 is a brief list of some audiovisual materials along with addresses where they can be obtained.

I am indebted to a large number of people for their assistance in the preparation of this book. It was my original intent to have Professor K. Read co-edit this book, but unforeseen events transpired which prevented his putting a serious effort into the program. My thanks go to all the contributors who cooperated with my ideas and recommendations, especially in the rewrite stages, to my fellow academicians who provided me with suggestions, to my wife, Ruth, who did the endless typing with patience and skill, and to the officers of Cambridge University Press, especially Florence Padgett of the North American office, with whom it is a distinct pleasure to work.

R. A. Granger

Introduction

Background

Engineering students are unique. They are usually uninterested in a problem unless they can visualize it. There are two ways visualization can be accomplished. One can create a mathematical model where mathematical symbols simulate properties, devices, and behaviors, or one can create the engineering problem in the laboratory. The former is usually faster and easier for the teacher, and the latter appears to be disappearing from educational institutions due to the ease and familiarity with the computer.

Motivation

The motivation behind this book is based on three quotations:

> The most effective method ever devised for teaching science – having students do experiments in a classroom laboratory to enable them to see the results – is slowly vanishing from American schools.
> *Boyce Rensberger, Washington Post, 11/12/88*

> Some schools have abandoned experiments in the lab in favor of simulated experiments on a computer that displays set-ups. This kind of thing is no substitute for a teacher or for a real lab.
> *George Tressel, Staff Associate of NFS's Education and*
> *Human Resources Division*

> I hear and I forget. I see and I remember. I do and I understand.
> *Old Chinese proverb*

Surely excellence in instruction is at the very root of education, and of necessity demands the maintenance of good academic standards. In that light, it follows that performing experiments is the grist of engineering. Nothing can be more significant than the marriage of excellent instruction incorporating well-defined academic standards with student involvement in the laboratory, that is, having the student put that instruction to practical use. What better way is there to insure the success of this idea than having the student perform laboratory experiments that are the favorite experiments of leading academicians?

Objectives

This book was conceived with three objectives in mind:

Provide a supplementary text that has more practical engineering as shown through the laboratory than through a single text having a collection of equations and sketches (customarily used in a theoretical course).

Present thermodynamics more in the image in which it was founded, that is, emphasize the physics of the phenomena rather than the mathematical model.

Make heat transfer practical as well as theoretical.

Each of the contributed experiments adheres to a similar format. Hence, the student is exposed to a scientific deductive reasoning process that can be used in any engineering course, research, or industry. The reasoning process is cycled over and over again from one experiment to the next, exposing the student to a manner of reasoning whether conscious of it or not.

The impact of a reasoning process on engineering education is enormous. The reasoning process usually involves free-body diagrams, control-volume sketches, and so on, each incorporating a visualization of the problem. Engineers are visual people. One of the ills in teaching thermo and heat transfer solely by classroom lecture is that the visual aspect of the problem is often omitted, scanty, or unclear. The presentation of engineering topics via the laboratory is one of the most visual of all approaches in presenting the subject.

General approach

This collection of experiments is intended to build on the background, preparation, and experience of the student.

We suggest that each lesson, each experiment, be personally supervised in practically every step of the procedure. It is recognized every student does not have the practical experience to anticipate a following step. Furthermore, we urge that the professor be present to oversee and explain equipment adjustments, calibrations, and instrumentation accuracy. The presence of the faculty during the conduct of the experiments should contribute to a lively and interesting session.

For both the thermo and heat-transfer laboratories, we recommend that students work in groups of 2 or 3. The faculty and technician should circulate throughout the lab and talk to individual student groups as the need arises. In each of the contributed experiments, short "mini-lectures" covering the background section of the experiment could enlighten the experiment.

Because experiments for the two courses are the favorites of the contributors, it is essential that each faculty member conducting the experiment preserve the spirit of that experiment; namely, try to make it the favorite of the student. Certainly problems will and should arise during the performance of the experiment, but problems that appear overwhelming to the student may be comprehensible to the group. Students working in small groups obtain training in shared information processing, which usually results in a greater comprehension of a problem than working in large groups.

Format

The format the contributors were asked to adhere to consisted of 9 key parts.

PRINCIPLE. What underlying principle is being stressed?

OBJECT. What does the experiment demonstrate?

BACKGROUND. A theoretical or physical explanation is presented that contains all equations to make the necessary calculations. This section also contains the rationale behind the experiment.

APPARATUS. A description of all apparatus and instrumentation necessary to conduct the experiment is presented. The apparatus is nonspecific and easily built or obtainable.

PROCEDURE. A step-by-step procedure is clearly presented so a novice can perform the steps.

SUGGESTED HEADINGS. A tabular form of quantities to be measured and calculated is given. Where necessary, typical results, figures, and sketches are presented to clarify the quantities to be measured.

REFERENCES. A collection of sources for students to obtain in-depth explanations is presented.

NOTATION. A listing of unusual symbols that may not be common or may have several meanings is sometimes given.

VITA. A very brief biography of the contributor plus his photograph is presented so the reader may appreciate the source of the experiment.

In addition, a DISCUSSION could be added if the contributor believed explanations had to be presented to clarify specific points.

Measurement uncertainty

Every experimental measurement carries error with it. Every engineer who designs the measurement system makes an estimate of the largest error that is expected to remain with the experiment. This estimate is called the *uncertainty*. For the sake of brevity, the method of measurement uncertainty is not presented herein. Appendix C of the text *Experiments in Fluid Mechanics** presents the uncertainty analysis in considerable detail. Examples are presented to show how this method of error analysis works.

* *Experiments in Fluid Mechanics*, ed. Robert A. Granger, Holt, Rinehart, and Winston, New York, 1988.

Concluding remark

When we shortchange our children in education, we rob them of their future, and cloud the country's as well. We must not only provide them with the best tools for learning but far more important and more difficult, instill in them the *desire to learn*. If we fail in this effort, considerations about the quality of a particular advanced curriculum become moot.

Elaine J. Camhi, Editor, Aerospace America, April 1992

Fig. 0.1. Photograph of laboratory set up to calibrate a thermometer near the ice point. (P. Nathanson student project. Photographed by D. B. Eckard, USNA)

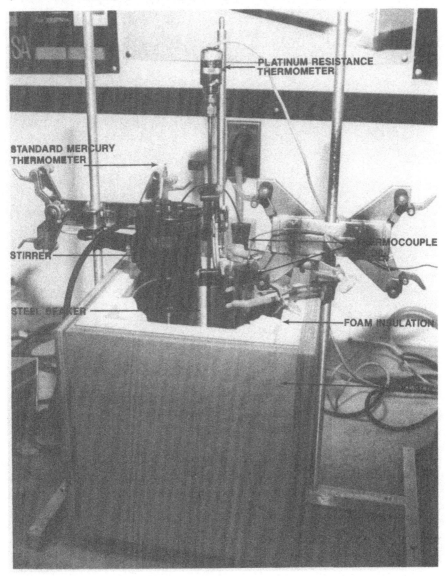

PART I
Experiments in heat transfer

We talk of heat as energy in the *process* of being transferred. Note, heat is *not* stored within matter, but rather heat is either "done on" or "done by" matter. Heat is a way of transferring energy across the boundaries of a system. It should be noted that heat is not a conserved substance, as was thought in the past (a remnant of the caloric theory of heat). Also, it is not a fluid as one might envision when the phrase heat flow is used, nor is heat a property of matter. Thus terms such as "heat of a substance" are meaningless. Since heat is neither a property of a system nor contained in a system, we speak of heat as a mode of energy transfer accompanied by a net amount of entropy transfer uniquely specified by the energy transfer as well as the temperature at which it occurs.

We may transfer heat by three different modes: conduction, convection, and radiation. Since each mode is subject to different laws, experiments such as those contained herein are necessary in order to understand the physical aspects involved in a heat-transfer problem.

PART I.1
Conduction

Fig. I.1. Typical microstructure of steel after exposure to a thermal cycle (magnification 250x). Zone 1 is the heat-affected region. Zone 2 is martensite unaffected by heat. (Courtesy of A. R. Imam and R. W. Haskell, *14th Conf. Thermal Conductivity*, Univ. CT, 1975.)

EXPERIMENT 1
Critical radius of insulation

Contributed by
WARREN M. ROHSENOW

Principle

Putting insulation on a small-diameter cylinder (or sphere) can increase the heat transfer. The diameter of the cylinder (or sphere) must be above a critical radius ($r_{crit.} = k_1/h$ for a cylinder) before insulation will reduce the heat transfer. This is why insulated electric wires can withstand more current than uninsulated ones. At the same wire temperature, insulated wires transfer more heat. Small- versus large-diameter hot-water pipes experience the same phenomenon.

Object

This classroom experiment is designed to demonstrate clearly the effect of small wire radius on insulated- and uninsulated-wire heat transfer rate.

Background

The heat transfer rate q from the inner surface of a tube to the surrounding air is

$$\frac{q/L}{2\pi(T_i - T_a)} = \frac{1}{\dfrac{1}{hr_o} + \dfrac{1}{k_g}\ln\dfrac{r_o}{r_i}} \tag{1.1}$$

where Fig. 1.1 shows r_i and r_o to be the inner and outer radii of a tube, respectively, and T_i and T_a the inner and outer air absolute temperatures, respectively. If all quantities are constant except for the expression on the left-hand side of Eq. (1.1) and r_o, differentiating Eq. (1.1) with respect to r_o results in the critical radius

$$r_{o,crit.} = k_g/h \tag{1.2}$$

5

Equation (1.2) is the condition for $(q/L)/(T_i - T_a)$ to be a maximum since the second derivative of Eq. (1.1) is negative.

Apparatus

Small-diameter electric wire
Glass tube of i.d. slightly larger than the electric wire, and o.d. approximately
    3–5 mm
Insulated supports for the wire to be held horizontally with the glass tube
    occupying half the central portion of the wire
Variac to vary the current in the wire.

  The experimental setup is shown in Fig. 1.2.

Fig. 1.1. Geometry for a tube.

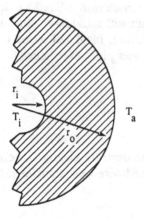

Fig. 1.2. Apparatus.

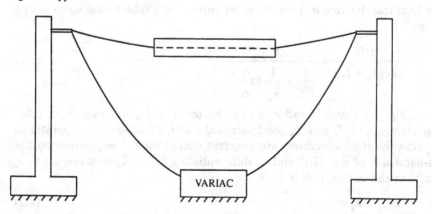

Procedure

Turn up variac until bare wire glows red or orange. (The portion of the wire
  in the glass tube will be dark colored indicating a lower temperature
  than the exposed wire.)
Since i^2R (hence heat rate q/A) is essentially the same in both the exposed
  and glass covered portion of the wire, the thermal resistance from
  the wire through the glass tube to air is less than from the uninsulated
  wire to air since it requires less ΔT to transfer essentially the same
  heat.

Discussion

Figure 1.3 is a plot of Eq. (1.1) for $k_g/h = 1.0$. Note at point A with $r_i = 0.50$
mm, adding insulation increases $q/\Delta T$ until r_o is 3.0 mm or greater, which is
a waste of insulation.

This experiment* demonstrates why insulated electric wires can carry more

Fig. 1.3. Effect of insulation thickness on heat loss from tubes showing critical radius of
insulation.

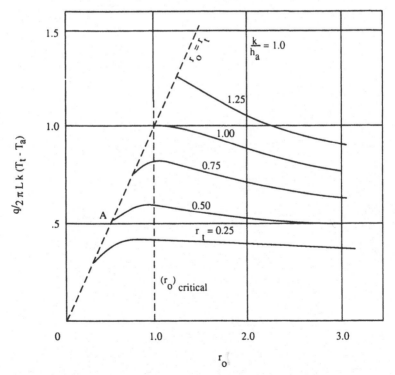

* This apparatus was originally built by Professor Gordon B. Wilkes of the M.I.T. Heat Transfer
Lab.

current (i^2R) than uninsulated wires. Also, hot-water or steam pipes less than around 0.75 in. should not be insulated to reduce heat loss. However, cold-water pipes should be insulated to prevent condensation and water damage.

As a challenging problem, have the students show that $h \propto \Delta T^{1/4}$. To make the experiment more interesting, give the students the glass tube and wire dimensions along with the current i and have them calculate T_{wire} bare and insulated.

Warren M. Rohsenow

Warren Rohsenow was educated at Northwestern Univ. (B.S. in mechanical engineering) and Yale Univ. (M. Eng., D. Eng.) and spent two years as an officer in the USN at the Engineering Experimental Station in Annapolis, MD. Since 1946, he has been at MIT as a professor and director of the Heat Transfer Laboratory. On September 25, 1992, MIT dedicated and named his lab the Warren M. Rohensow Heat and Mass Transfer Laboratory. He is also co-founder of Dynatech Corp.

EXPERIMENT 2
The regelation of ice – the effect of heat conduction

Contributed by
ERICH W. P. HAHNE

Principle

Materials that contract while melting have a melting temperature (melting point) that decreases when pressure increases. Such materials, in solid state, swim on their melt and exhibit the phenomenon of regelation.

Object

In some physics textbooks or physics lectures the following experiment is presented: A wire loop, slung around an ice block and loaded with a weight, penetrates through the block without cutting it apart. The effect, which sounds like magic, already puzzled Michael Faraday and John Tyndall over 100 years ago. John Tyndall named it "regelation" (re + *gelare*, Lat. refreezing).

The usual explanation in textbooks is

> The ice melts under the increased pressure of the weighted wire and the water freezes again when the pressure is released above the wire.

This explanation is certainly true, as from the Clausius–Clapeyron equation applied to the melting of ice, we obtain

$$\Delta T = -0.0074 \, \Delta p. \tag{2.1}$$

This means that an increase in pressure by, for example, $\Delta p = 1$ bar results in a change of melting temperature of $\Delta T = -0.0074$ K. But is pressure the only parameter which affects this phenomenon? If it were so, the material of the wire, whether metal or plastic, should not have any effect on the velocity of the wire penetrating through the ice.

Apparatus

For the simple experiment, equipment can be used which is in every workshop or can be made easily:

A rectangular trough ca. $20 \times 10 \times 10$ cm³ for the preparation of an ice block
Two U-shaped wooden frames
Two weights
A copper wire and a nylon string of diameter 0.5 mm, length 15 cm
Two supports; pieces of styrofoam
A watch; a ruler

The experimental setup, which can be in any room of about 20 °C, is shown in Fig. 2.1.

The ice block is prepared in the trough and will have a volume of about $20 \times 10 \times 10$ cm³. It can be longer and wider, but should not be much smaller. Special care should be taken in making the ice block. This will be explained in more detail in a later section.

The ice block rests on styrofoam so that the melting rate on both supporting ends is decreased. The U-frames should be about the width of the ice block, so that the wires on both sides have only little contact area with the ambient temperature and heat flow into the ice is small. The frames must have an indicator (e.g., nail) on both sides at the same height. The wire or string spanned between the legs of the frame should be as tight as possible so that its curvature in the ice is small. Thus, only a little water escapes from the refreezing zone and the pressure exerted by the wire upon the ice is better defined. One frame will be equipped with the copper wire, the other with the nylon string. The frames are slipped over the ice blocks and loaded with the weights.

With the ruler and the indicators, the penetration distance Δl can be measured and with the respective time interval Δt a penetration velocity

$$w = \Delta l / \Delta t \qquad (2.2)$$

can be calculated.

The experiments will yield clear results; but their reproducibility is poor,

Fig. 2.1. Experimental set up.

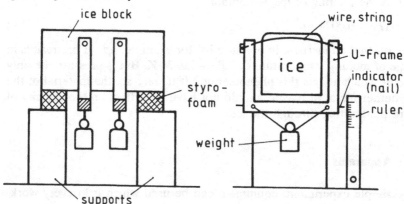

that is, the results scatter in a wide band when repeated experiments are performed under presumed equal conditions. The reason is the widely varying quality of ice.

Procedure

If we take water from the faucet and have it frozen in the refrigerator, we obtain ice with a high content of air. This ice is opaque with more or less air here or there. The distribution and amount differ from one block to the other. A criterion for ice blocks of comparable quality is their transparency: With no or very little air the ice is transparent. Such a transparent ice block is shown in Fig. 2.2. Behind this block is a black photocardboard with the word "ICE" cut into it. This kind of ice is obtained from distilled water that was boiled for about 15 minutes and cooled in a –5 °C brine bath under continuous stirring with a motor-mixer. Frequent stirring is necessary to drive out the air when the ice is prepared in a refrigerator.

The readings of the penetration distance Δl and the respective penetration time interval can be listed in a table as shown in Fig. 2.3. The pressure exerted by the wire is calculated from

$$p = \frac{\text{weight}}{\text{wire area}} = \frac{\text{weight}}{d \cdot l_{\text{ice}}} \tag{2.3}$$

with d being the wire diameter and l_{ice} the length of the wire within the ice. Equation (2.3) assumes that the wire in the ice is horizontal.

Fig. 2.2. Transparent ice block.

Results

The result for a copper wire and a nylon string is undoubtedly clear: The copper wire is more than six times faster than the nylon string although pressure and diameter are the same.

The experiment may be repeated with other materials, for example, an iron wire or a silk string. The iron wire will penetrate slower than copper, but much faster than silk, which is a little slower than nylon.

Explanation

This large difference of penetration velocity can be explained by the large difference of thermal conductivity: Copper, with $k \approx 350$ W/K·m is an excellent thermal conductor whereas nylon with $k \approx 0.3$ W/K·m is a poor conductor. Why does this make a difference? When the ice melts under the wire, heat of melting is required there; when the water above the wire refreezes, heat of melting is liberated there. From the heat source on top of the wire the heat of melting will flow to the heat sink at the bottom of the wire. This flow can be through the wire or around it, through the ice, wherever it finds the least resistance. The thermal resistance depends on the length of the heat flow lines – it is larger, when the flow lines are longer – and on the thermal conductivity – the resistance is large, when the conductivity is small. With the

Fig. 2.3. Example of a list of experimental results.

Material	diameter d mm	pressure p bar	penetration distance Δl mm	penetration time interval Δt min	penetration velocity w mm/min
Copper	0.5	10	6	5	1.2
			11	10	1.1
			10	10	1.0
			11	10	1.1
			$\bar{w}$ = 1.1 mm/min ≈ 66 mm/h		
Nylon	0.5	10	3	20	0.15
			4	20	0.20
			3.5	20	0.175
			2.5	20	0.125
			$\bar{w}$ = 0.1625 mm/min ≈ 9.75 mm/h		

copper wire most of the heat will flow through the wire; with the nylon string most of the heat will flow through the ice ($k \approx 2$ W/K · m). This regelation model is shown in Fig. 2.4.

There is – and this makes a theoretical approach very complicated – a water layer ($k \approx 0.6$ W/K · m) around the penetrating material. In order to calculate its thermal resistance the width of this layer has to be known. But we have to rely on guesses; various authors assume values between 0.2 and 1 mm. Several theoretical approaches have been made, but theory and experiment still differ by some 100 percent.

Besides the water layer thickness, there are other open questions: Is there an interaction between the penetrating material and the ice; is there an effect of surface roughness; how homogeneous is the ice; does all the water refreeze again? As far as the homogeneity is concerned Fig. 2.5 shows the transparent ice block filled with a network of little channels when it is treated with a

Fig. 2.4. Heat-conduction regelation model.

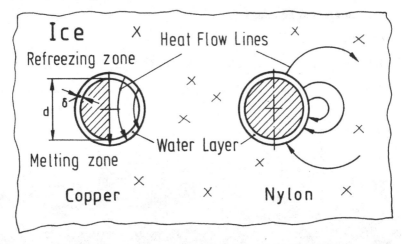

Fig. 2.5. Transparent ice block with fluorescent liquid along the crystal boundaries.

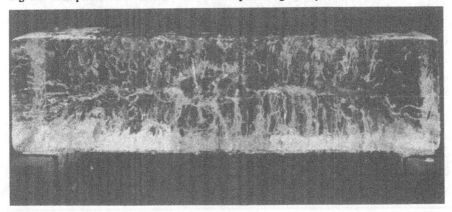

solution of sodium fluorescein ($C_{20}H_{10}Na_2O_5$) in water. If such a solution is poured on the block it is siphoned into the ice along the crystal boundaries. Under an ultraviolet (UV) lamp only the liquid solution becomes visible; the solid ice does not absorb fluorescein. Thus the theory of a homogeneous ice state could only hold for a monocrystal, not for multicrystal ice blocks where liquid interfaces between crystal boundaries, which are arbitrarily located, disturb the heat flow field.

Another photograph, Fig. 2.6, shows the wire while it is penetrating through the ice: Again the fluorescein solution indicates, as a yellow line, the liquid layer around the wire. The upper part of the picture shows the refrozen zone, where yellow spots indicate unfrozen liquid islands.

A simple experiment, performed with modest, unsophisticated equipment reveals a series of unexpected problems. By the experiment we obtain a clear, logical result, but we cannot predict it quantitatively. An old problem, which on first sight quite simple, still waits for a solution.

Suggested headings

Constants:

Material 1: _____   Material 2: _____

Weight: _____   Weight: _____

Fig. 2.6. Wire penetration through the ice.

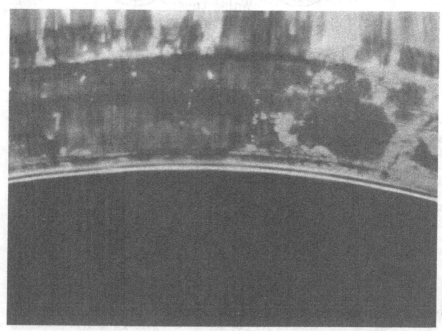

Material diameter: _____   Material diameter: _____

Length in ice: _____   Length in ice: _____

Pressure: _____   Pressure: _____

Material 1					Material 2				
l_1	l_2	Δl	Δt	w	l_1	l_2	Δl	Δt	w

Erich W. P. Hahne

Erich Hahne was educated at the Technical University München (D. Eng.). With a NASA research grant he spent a year at the California Institute of Technology, Pasadena. Since 1973 he has been a professor and holder of the Chair for Thermodynamics and Heat Transfer at the Technical University Stuttgart. His main research fields are now boiling heat transfer, high-temperature thermal properties, and solar-energy applications.

EXPERIMENT 3
Unsteady heat conduction in a sphere

Contributed by
S. G. BANKOFF

Principle

One method of measuring thermal diffusivities of different solids is to immerse a sphere of the material in a hot (or cold) water bath, and to measure the temperature response at different points within the solid.

Objective

To provide a simple undergraduate experiment illustrating Fourier's Law for unsteady heat conduction.

Apparatus

A 5–8-cm-diameter sphere is used of the material whose thermal conductivity it is desired to measure. Plexiglas is good because one can see the placement of the thermocouples. Other possible choices include wood, rubber, sponge rubber, or even such familiar objects as an apple or an orange. For soft materials sheathed, hypodermic-type thermocouples are preferable. For hard materials small radial holes are drilled to the center and to the midradius. Thermocouples are inserted into the bottom of these holes, and good thermal contact is ensured by using conductive heat transfer paste. The holes are sealed against water entry by silicone or other sealant. A constant-temperature water bath with a stirrer is required, together with a frame to hold the sphere. In the simplest version the thermocouples are read manually from a potentiometer with a thermocouple switch, or from two potentiometers. A preferable arrangement uses amplifiers for the two inserted thermocouples, with digital or analog readout. A multichannel temperature scanner may be used, or the instructor may wish to write a simple program for sampling the data and storing in a PC. A thermocouple in the water bath is also monitored to ensure constant temperature conditions.

16

Procedure

The water bath is brought to a constant temperature of 60–5 °C, and the sample holder is suddenly immersed. Readings are taken as often as possible. After the sphere has come to temperature equilibrium, it may be immersed in an ice-water bath and the process repeated. The Biot number may also be varied by running with the stirrer on and off.

The one-term approximation to the exact solution is

$$\theta^* = C_1 \exp(-\zeta_1^2 Fo)\, \frac{1}{\zeta_1 r^*} \sin(\zeta_1 r^*) \tag{3.1}$$

or $\qquad \theta^* = \theta_0^* \dfrac{1}{\zeta_1 r^*} \sin(\zeta_1 r^*) \tag{3.2}$

where

$$\theta^* = \frac{T - T_\infty}{T_1 - T_\infty}$$

is the dimensionless temperature at $r^* = \dfrac{r}{r_0}$, and

$$\theta_0^* = \frac{T_0 - T_\infty}{T_1 - T_\infty}$$

is the dimensionless center temperature, given by

$$\theta_0^* = C_1 \exp(-\zeta_1^2 Fo) \tag{3.3}$$

where

$\qquad T_\infty$ = liquid temperature
$\qquad T_1$ = initial temperature
$\quad T(r,t)$ = temperature at radial position r and time t
$\qquad r_0$ = sphere radius
$\qquad Fo = \dfrac{\alpha t}{r_0^2}$ Fourier number
$\qquad \alpha$ = thermal diffusivity of solid
$\qquad Bi = \dfrac{hr_0}{k}$ = Biot number
$\qquad h$ = surface heat transfer coefficient
$\qquad k$ = thermal conductivity of solid

C_1 and ζ_1 are functions of Bi, and are tabulated in Incropera and Dewitt, *Fundamentals of Heat and Mass Transfer*, Chap. 5, as well as other texts. This approximation holds for $Fo \geq 0.2$.

Thus, a semilog plot of θ_0^* versus time should give a straight line whose slope and intercept with $t = 0$ give $\dfrac{-\zeta_1^2 \alpha}{r_0^2}$ and C_1, respectively. From the table, one can thus determine Bi, and thence ζ_1 and α. For smaller Fourier numbers the full series solution should be used. In this case the one-term solution should be used to give initial values for *Bi* and α, which are then adjusted iteratively to give a good fit to the data.

To check your calculations, the midradius temperature data can be used, together with Eq. (3.2). Another check can be made with the graphical (Heisler) charts found in nearly all heat transfer texts. The student should tabulate the values of α and *Bi* thus found, and discuss the probable reasons for any discrepancies.

Acknowledgment: This experiment, which is used in the undergraduate chemical engineering laboratory course at Northwestern, was designed and put into operation by Prof. J. S. Dranoff.

Suggested headings

Constants:

$$T_1 = \qquad T_\infty = \qquad r_0 = \qquad r_{1/2} =$$

$$T_0 \qquad\qquad T_{1/2} \qquad\quad t \qquad\quad \theta_0^* \qquad \theta_{1/2}^*$$

S. George Bankoff

S. George Bankoff is Walter P. Murphy Professor Emeritus of Chemical and Mechanical Engineering at Northwestern University in Evanston, IL. and Director of the Center for Multiphase Flow and Transport. He is the winner of the 1987 Max Jakob Memorial Award in Heat transfer.

EXPERIMENT 4
Heat conduction in materials with nonhomogeneous structure

Contributed by
WLADYSLAW KAMINSKI

Principle

A linear heat source placed in a material of semi-infinite nature causes temperature changes due to heat conduction. The relationship between temperature at a given distance from the heat source and time in the coordinate system $(T - T_o, \log t)$ is linear after some time.

Object

The experiment demonstrates heat conduction in semi-infinite space. It allows one also to obtain experimentally both the thermal conductivity and the thermal diffusivity of the material. An object used for the experiment can be any material in the form of powder, granules, or paste in bulk.

Apparatus

The apparatus consists of a glass beaker (diameter 0.1 m, height 0.2 m) which is filled with a selected material. At the center of the beaker a linear heat source 0.1 m long is placed and there is a temperature sensor parallel to the heater at a distance of 10 mm from the heater (Fig. 4.1). Both of them are in the form of a capillary metal tube 0.2–0.5 mm in diameter. The heat source is connected to a stabilized power supply (approximately 1.5 W) and the temperature sensor to a display and/or to a recorder.

Procedure

The experiment is carried out at ambient temperature. Switch on the power and register temperature in 2 minute intervals for 20 minutes. After completing the experiment remove the heat source and the temperature sensor to cool them to ambient temperature. Mark experimental points in the coordinate system ($y = T - T_o$, $x = \log t$) and determine the moment the relationship becomes linear. Calculate the constants C_1 and C_2 in the equation $y = C_1 x + C_2$ using the least squares method and then the thermal conductivity and thermal diffusivity.

Explanation

Consider a linear heat source of capacity Q_1 [J/m] in nonhomogeneous materials of bulk density ρ_n, heat capacity c_p, and thermal diffusivity a. The initial temperature of the material is T_o. Heat transfer from the heater to the material is due to the cumulative effect of different transfer mechanisms. This heat-transfer process is described by Fourier's equation

$$q = -k \, \nabla T \tag{4.1}$$

where k is the thermal conductivity for homogeneous materials but here represents a cumulative effect of heat transfer. The heat transfer from the heater to a material causes temperature changes. Temperature at a distance r from the heater at a moment of time t can be described by Eq. (4.2).

$$T - T_o = \frac{Q_1(t)}{4\pi k t} \exp\left(-\frac{r^2 + r_o^2}{4at}\right) I_o\left(\frac{rr_o}{2at}\right) \tag{4.2}$$

To get a useful form of Eq. (4.2) some simplifications should be considered which can be achieved easily in an experiment:

Fig. 4.1. Scheme of experimental setup:
1. To power supply and stabilizer. 2. To display and/or recorder. 3. Support. 4. Temperature sensor. 5. Heat source. 5.1. Wall. 5.2. Electric isolation. 5.3. Resistance wire. 6. Material examined. 7. Glass beaker.

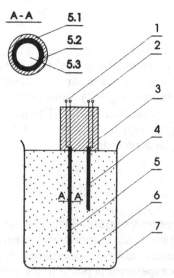

If $\quad r_o \to 0$ and $t \to \infty$ then $I_o\left(\dfrac{rr_o}{2at}\right) \to 1$ $\qquad$ (4.3)

and $\quad Q_1(t) = $ const. $\qquad$ (4.4)

If we take into account assumptions (4.3) and (4.4) Eq. (4.2) becomes

$$T - T_o = \frac{q_1}{4\pi k}\left[\log t + \log\left(\frac{4a}{r^2}\right) - \gamma\right] \qquad (4.5)$$

where $\gamma = 0.5772\ldots$ is Euler's constant and q_1 is the heat-source power (in W/m). It means that Eq. (4.5) under assumptions (4.3) and (4.4) in the co-ordinate system $y = T - T_o$, $x = \log t$ can be presented in the form of a linear equation

$$y = C_1 x + C_2 \qquad (4.6)$$

On the basis of experiments the constants C_1 and C_2 can be calculated. It follows from Eq. (4.5) that C_1 and C_2 can be expressed as:

$$C_1 = \frac{q_1}{4\pi k} \qquad (4.7)$$

$$C_2 = C_1\left[\log\left(\frac{4a}{r^2}\right) - \gamma\right] \qquad (4.8)$$

Suggested headings

Constants: $q_1 = $ _____ ; $r = $ _____

Time(s)	log t	$T - T_o$

References

1. Carslaw, H. S., and Jaegar, J. C., *Conduction of Heat in Solids*, Oxford University Press, London, 1959.
2. Ohotin, A. C., "Teploprovodnost tverdykh tel," *Gos. Izd. Tekhn. Teor. Liter.*, Moskva, 1984 (in Russian).
3. Schulte, K., "Instationäre Mezsonden zur Bestimmung der Wärmeleitfähigkeit und der Wärmekapazität von Feststoffen und Schüttgutern," *Energietechnik* 36, 11 (1986): 416–19.

22   W. Kaminski

Notation

a	m²/s	thermal diffusivity
C_1, C_2		constants
I_o		zeroth-order modified Bessel function of first kind
k	W/(m K)	thermal conductivity
log		logarithm at natural base
q	W/m²	heat flux
q_1	W/m	power of heat source
Q_1	J/m	heat source capacity
r	m	distance
r_o	m	radius of linear heat source
t	s	time
T	°C	temperature
T_o	°C	initial temperature

Wladyslaw Kaminski

Professor Kaminski received his Ph.D. at the Technical University at Lodz and his D.Sc. at the Technical University of Silesia, both in chemical engineering. He has conducted numerous research projects in heat convection, specializing in the drying of biosynthetic products and spouted and fluidized bed processes.

EXPERIMENT 5

Measurement of thermal conductivity of solids during chemical reactions

Contributed by

H. MATSUDA *and* M. HASATANI

Objective

Thermal conductivity is one of the most important thermophysical properties in the evaluation of heat flow rate within a solid. Various methods such as the absolute method, the twin plate method, the hot-wire method,[9,10] and so forth, so far have been proposed for the thermal conductivity measurement of solids. These conventional methods, however, are not applicable for measurements of thermal conductivity when the materials to be measured are subject to phase change or chemical reaction. It is difficult in these methods to eliminate the additional heat flow released by phase change or reaction from the total heat flow.

A novel method for measuring thermal conductivity of solids in such an unsteady-state accompanied by heat-generation or heat-absorption is described. In this method, the value of the effective thermal conductivity in the process of reaction is evaluated from the integrated time change of the temperature of the reacting solid material (D.T.A. curve), by removing the effect of the reaction heat.

Background

The principle equation for determining the thermal conductivity in the process of reaction can be derived by referring to a D.T.A. (differential thermal analysis) measurement (see Table 5.1). The following assumptions are made: (1) Both the reference(I) and the reactant(II) samples are cylindrical; (2) The axial heat flow in the sample is negligible; (3) The reaction heat H is lumped[1-3] with specific heat c_p as:

$$C_p(t) = c_p + H\frac{\partial k}{\partial t}.$$

Under the assumptions (1) and (2), the fundamental equations with respect to heat conduction in each sample can be expressed as:

Reference sample:

23

$$c_{pI}\rho_{I}\left(\frac{\partial t_{I}}{\partial \theta}\right) = \frac{1}{r}\frac{\partial}{\partial r}\left(\lambda r \left(\frac{\partial t_{I}}{\partial r}\right)\right) \tag{5.1}$$

Reactant sample:

$$c_{pII}\rho_{II}\left(\frac{\partial t_{II}}{\partial \theta}\right) = \frac{1}{r}\frac{\partial}{\partial r}\left(\Lambda(t_{II})r\left(\frac{\partial t_{II}}{\partial r}\right)\right) - H\rho_{II}\left(\frac{\partial k}{\partial \theta}\right) \tag{5.2}$$

where $\Lambda(t_{II})$ denotes the thermal conductivity in the process of reaction; I.C. and B.C.:

$$\left.\begin{array}{l} 0 \leqq r \leqq a, \ \theta = \theta_0; \ t_I = t_{II} = t_0 \\ 0 \leqq r \leqq a, \ \theta = \theta_1; \ t_I = t_{II} = t_1 \\ \theta \geqq \theta_0, \ r = a; \ t_I = t_{II} = \gamma\theta + t_0 \ (\gamma \text{ constant}) \\ \theta \geqq \theta_0, \ r = 0; \ \partial t_I/\partial r = \partial t_{II}/\partial r = 0 \end{array}\right\} \tag{5.3}$$

Rewriting Eq. (5.2) by use of $C_p(t)$ in assumption (3), and introducing the following variables

$$\frac{c_p(t_{II})dt_{II}}{c_{pII}} = d\psi, \qquad \frac{\Lambda(t_{II})dt_{II}}{\lambda} = d\phi \tag{5.4}$$

we can rewrite Eq. (5.2) as

$$c_{pII}\rho_{II}\left(\frac{\partial \psi}{\partial \theta}\right) = \frac{1}{r}\frac{\partial}{\partial r}\left(\lambda r\left(\frac{\partial \phi}{\partial r}\right)\right) \tag{5.5}$$

Putting $c_p = c_{pI} = c_{pII}$, $\rho = \rho_I = \rho_{II}$ and assuming c_p, ρ, and λ to be constant, we can obtain Eq. (5.6) by subtracting Eq. (5.5) from Eq. (5.1).

Table 5.1. *Example of reactant sample*

Reactant material	Reaction type	Reaction temperature [K]	Heat of reaction [kJ · kg⁻¹]
$NaHCO_3$	dehydration & decarbonation	393.2	855.1
n.$CaSO_4 \cdot 2H_2O^a$	dehydration	403.2	489.3
n.$CaSO_4 \cdot 1/2H_2O^a$	↑	433.2	115.5
c.$CaSO_4 \cdot 2H_2O^a$	↑	388.2	489.3
c.$CaSO_4 \cdot 1/2H_2O^a$	↑	463.2	115.5
NH_4Cl	crystal transformation	453.2	78.3
AgI	↑	420.2	26.2
SiO_2	↑	846.2	10.5
$BaCO_3$	↑	1083.2	75.3
$Ca(OH)_2$	dehydration	793.2	1410.0
$CaCO_3$	decarbonation	1173.2	1780.0

ᵃ (n = natural, c = chemical)

$$c_p \rho \left(\frac{\partial \psi'}{\partial \theta} \right) = \frac{1}{r} \frac{\partial}{\partial r} \left(\lambda r \left(\frac{\partial \phi'}{\partial r} \right) \right) \tag{5.6}$$

where $\psi' = \psi - t_1$, $\phi' = \phi - t_1$.

By assuming $C_p(t_{II})/c_p$ and $\Lambda(t_{II})/\lambda$ are constant during reaction as $C_p(t_{II})/c_p = 1 + K_1$ and $\Lambda(t_{II})/\lambda = 1 + K_2$, Eq. (5.6) is rewritten as Eq. (5.7).

$$(1 + K_1) \frac{c_p \rho}{\lambda} \left(\frac{\partial T}{\partial \theta} \right) + \left(\frac{c_p \rho K_1}{\lambda} \right) \left(\frac{\partial t_1}{\partial \theta} \right) = \frac{(1 + K_2) \partial}{r} \frac{\partial}{\partial r} \left(r \frac{\partial T}{\partial r} \right) + \frac{K_2 \partial}{r} \frac{\partial}{\partial r} \left(r \frac{\partial t_1}{\partial r} \right) \tag{5.7}$$

where T represents the differential temperature between the reactant and the reference sample, $T = t_{II} - t_1$. The integration of Eq. (5.7) over time yields Eq. (5.8) by use of both the initial and the boundary conditions (5.3).

$$\frac{c_p \rho K_1 (t_1 - t_0)}{\lambda} = \frac{(1 + K_2)}{r} \frac{\partial}{\partial r} \left(r \frac{\partial}{\partial r} \int_{\theta_0}^{\theta_1} T d\theta \right) + \frac{K_2 \partial}{r} \frac{\partial}{\partial r} \left(r \frac{\partial}{\partial r} \int_{\theta_0}^{\theta_1} t_1 d\theta \right) \tag{5.8}$$

Multiplying Eq. (5.8) by r, integrating from 0 to r, dividing by r, and then integrating from 0 to a yields,

$$\int_0^a \left(\frac{Q}{2\lambda} \right) r \, dr = - (1 + K_2) \int_{\theta_0}^{\theta_1} T_c d\theta + K_2 \int_{\theta_0}^{\theta_1} (t_{1,s} - t_{1,c}) d\theta \tag{5.9}$$

where T_c represents the differential temperature at the center between the reference sample and the reactant. The subscripts 0 and 1 express the beginning and the end of the reaction, respectively, and $Q(= H\rho)$ is the heat of reaction per unit volume. Introducing $t_{1,s} - t_{1,c} = \Delta t_1$ into the second term of the right-hand side of Eq. (5.9) yields

Material	ρ[kg·m^{-3}]	$c_p(t)$ [kJ·kg^{-1}·K^{-1}]	$\lambda(t)$ [W·m^{-1}·K^{-1}]
NaHCO$_3$	1650	1.21 (393.2K)	0.848 ($\leftarrow$)
n.CaSO$_4$·2H$_2$O	2320	1.19 (403.2K)	1.045 ($\leftarrow$)
n.CaSO$_4$·1/2H$_2$O	1955	1.05 (433.2K)	0.604 ($\leftarrow$)
c.CaSO$_4$·2H$_2$O	1600	1.18 (388.2K)	0.430 ($\leftarrow$)
c.CaSO$_4$·1/2H$_2$O	1348	1.08 (463.2K)	0.325 ($\leftarrow$)
NH$_4$Cl	1450	2.07 (453.2K)	0.639 ($\leftarrow$)
AgI	5030	0.276 (400.2K)	0.188 ($\leftarrow$)
BaCO$_3$	3000	0.502 (403.2K)	0.697 ($\leftarrow$)
Ca(OH)$_2$	1650	1.34 (653.2K)	2.324 ($\leftarrow$)
CaCO$_3$	2700	1.05 (423.2K)	2.673 ($\leftarrow$)

$$K_2 = \left(\frac{Qa^2}{4\lambda} + \int_{\theta_0}^{\theta_1} T_c d\theta\right) \Big/ \left(-\int_{\theta_0}^{\theta_1} T_c d\theta + \Delta t_I(\theta_1 - \theta_0)\right)$$ (5.10)

The thermal conductivity in the process of reaction Λ can be thus determined by Eq. (5.10) from the peak area of the D.T.A. curve, the reaction heat, and the average temperature difference between the center and the surface of reference sample, Δt_I.

Apparatus

The details of the sample holder are shown in Fig. 5.1. The sample holder is made of a cylindrical stainless-steel block of 60 mm in diameter and 90 mm in height. The sample holder is equipped with two holes (8.0 mm in diameter and 30 mm in depth) in a symmetrical position from the center so as to be charged with both the reactant and the reference sample. Chromel-alumel thermocouples of 0.3 mm in diameter are used for measuring the temperatures of the reactant, reference sample, and sample holder.

Fig. 5.1. Details of the sample holder.

1. Stainless–steel
   sample holder

2. Reference chamber

3. Sample chamber

4. 0.3mm⌀ Chromel–Alumel
   thermocouples

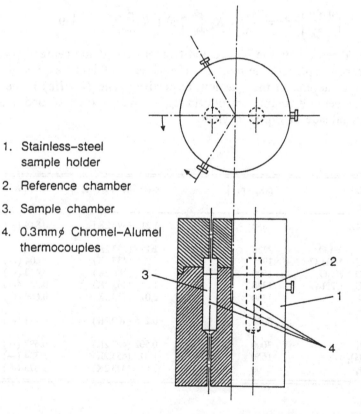

Procedure

The reactant and the reference samples are made in a cylindrical form (8.0 mm × 18–20 mm) by a press. They are equipped with a small wire hole along the center to set a thermocouple wire (see Fig. 5.1). Each thermocouple set in the reactant, the reference, and the sample holder is connected with a millivolt recorder and an automatic temperature recorder (see Fig. 5.2). Then, the sample holder is placed in the heating apparatus, and is heated at a constant rate.

The differential temperature T_c between the reactant and the reference sample is recorded continuously by a millivolt recorder. The average temperature difference Δt_l is also measured with another differential thermocouple. For the exothermic step in crystal transformation, the sample holder is cooled down at a constant rate in the heating apparatus after the completion of

Fig. 5.2. Schematic drawing of the experimental apparatus.

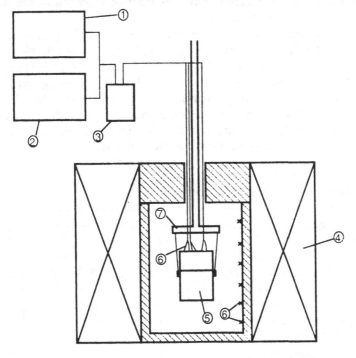

① Temperature recorder	⑤ Sample holder
② Millivolt recorder	⑥ Thermocouples
③ Cold junction	⑦ Supporter
④ Electric furnace	

the heating experiment, and the D.T.A. curve is obtained from this cooling experiment.

The ratios of the measured thermal conductivity in the process of reaction to that in the state before reaction, Λ/λ, are plotted in Fig. 5.3 against the heat of reaction $Q(=H\rho)$. As can be seen from the figure, there is observed a certain relation between Λ/λ and Q; the value of Λ/λ decreases with an increase in the heat of reaction. When the reaction heat is small enough, such as for the crystal transformation of quartz sand ($H = 10.5$ kJ kg^{-1}), there is no appreciable change in the thermal conductivity value. Whereas when the reaction heat becomes larger, the value of Λ/λ decreases and finally reaches an asymptotic value of about 0.25. It is difficult to explain exactly the reason why the value of Λ becomes smaller than that of λ, and why the ratio of Λ/λ is closely related to the reaction heat. As a qualitative interpretation of this thermal-conductivity behavior in the process of reaction, it is supposed that in the process of reaction the thermal wave that propagates through the solid may be disturbed by the tentative thermal vibration of the crystal structure caused by the reaction. In such a case, the heat of reaction is considered to be a measure of the magnitude of this thermal vibration of the bonded molecules in the crystal structure.

Typical examples of the D.T.A. curves for the crystal transformation of ammonium chloride and the thermal decomposition of sodium bicarbonate and gypsum (chemical and natural) are shown in Figs. 5.4 and 5.5. The chained lines in the figures are the D.T.A. curves calculated by using the thermal conductivity measured by the present method. It may be said that the measured thermal conductivity values are valid enough to reproduce the D.T.A. curves obtained in this experiment under various conditions. See Refs. 4–8 for details in the measured thermal conductivity data.

Fig. 5.3. Relation between Λ/λ and heat of reaction.

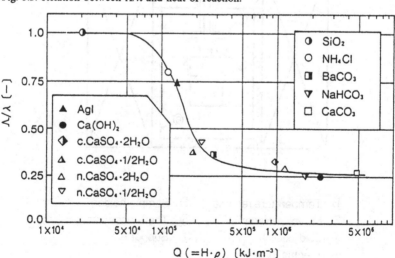

Suggested headings

Data:

$$a = \underline{\hspace{2cm}} ; \qquad H^a = \underline{\hspace{2cm}}$$

$$\rho = \underline{\hspace{2cm}} ; \qquad \lambda^b = \underline{\hspace{2cm}}$$

Time	Differential temperature	
	T_c	$t_{I,s} - t_{I,c}$
θ_0	_____	_____
θ_α	_____	_____
θ_β	_____	_____
θ_γ	_____	_____
.	.	.
.	.	.
.	.	.
θ_1	_____	_____
	$\displaystyle \lim_{\Delta\theta \to 0} \sum_{i=0}^{1} T_{ci}\Delta\theta$ $= \underline{\hspace{2cm}}$	$\displaystyle \Delta t_I = \frac{\sum_{i=0}^{1}(t_{I,s} - t_{I,c})_i \theta_i}{\sum_{i=0}^{1}\theta_i}$ $\Delta t_I(\theta_1 - \theta_0) = \underline{\hspace{2cm}}$

$$K_2 \left(= \frac{\Lambda - \lambda}{\lambda} \right) = \underline{\hspace{2cm}}$$

References

1. Hasatani, M., and Sugiyama, S., "Studies on two-stage thermal decomposition of solid," *Kagaku Kogaku* 28 (1964): 355–61.
2. Kato, T.; Kito, M.; Nakamura, M., and Sugiyama, S., "The effect of heat transfer on the rate of crystal transformation," *Kogyo Kagaku Zasshi* 68 (1965): 94–7.
3. Kito, M., and Sugiyama, S., "Thermal decomposition of sodium bicarbonate," *Kagaku Kogaku* 28 (1964): 814–19.
4. Matsuda, H., and Hasatani, M., "A consideration of the effect of phonon on thermal conductivity during the course of reaction state," *J. Chem. Eng., Japan* 8 (1985): 84–7.
5. Matsuda, H., and Hasatani, M., "Measurement of effective thermal conductivity of solids in the process of chemical reaction," *Experimental Heat Transfer, Fluid Mechanics and Thermodynamics* (1988): 1258–65.

[a] Refer to Table 5.1.
[b] Measured by conventional thermal conductivity measurement method (e.g., Nagasaka et al.[9] and Ven der Held and Van Drunan[10]); refer to Table 5.1.

6. Matsuda, H.; Hasatani, M., and Sugiyama, S., "Thermal conductivities of solids under crystal transformation and thermal decomposition reaction," *Kagaku Kogaku Ronbunshu* 1 (1975): 589–93.
7. Matsuda, H.; Hasatani, M., and Sugiyama, S., "Effects of heat of reaction on thermal conductivities of solids under reaction," *Kagaku Kogaku Ronbunshu* 2 (1976): 630–2.
8. Mogilevsky, B. M., and Chundnovsky, A. P., "The study of thermal conductivity of semiconductors in solid and melted states at high temperatures," *Proc. Int. Conf. Phys. Semicond.* 9th 2 (1968): 1241–5.
9. Nagasaka, K.; Shimizu, M., and Sugiyama, S., "End effect in thermal conductivity measurement by the transient method," *J. Chem. Japan* 6 (1973): 264–8.
10. Ven der Held, E. F. M., and Van Drunen, F. G., "A method of measuring the thermal conductivity of liquids," *Physicia* 15 (1949): 865–81.

Notation

a	radius of a sample	[m]
c_p	specific heat	[kJ kg^{-1} K^{-1}]
$C_p(t)$	overall specific heat, $C_p(t) = c_p + H\dfrac{\partial k}{\partial t}$	[kJ kg^{-1} K^{-1}]
H	heat of reaction per unit mass	[kJ kg^{-1}]

Fig. 5.4. D.T.A. curves for thermal decomposition of sodium bicarbonate and gypsum (natural and chemical).

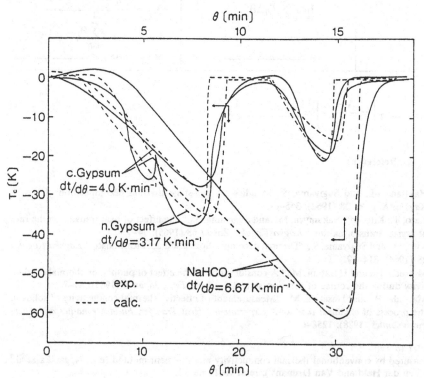

K_1	constant	[–]
K_2	constant	[–]
k	fraction of reaction	[–]
Q	heat of reaction per unit volume	[kJ m^{-3}]
r	distance from the center of a sample	[m]
T	differential temperature	[K]
t	temperature	[K]
Δt	temperature difference	[K]
θ	time	[s]
Λ	effective thermal conductivity in the process of reaction	[W m^{-1} K^{-1}]
λ	effective thermal conductivity of a reactant	[W m^{-1} K^{-1}]
ρ	bulk density of a sample solid	[kg m^{-3}]

Subscripts

I	reference
II	reactant
0	initial
1	final
c	center
s	surface

Fig. 5.5. D.T.A. curves for thermal decomposition of sodium bicarbonate and crystal transformation of ammonium chloride.

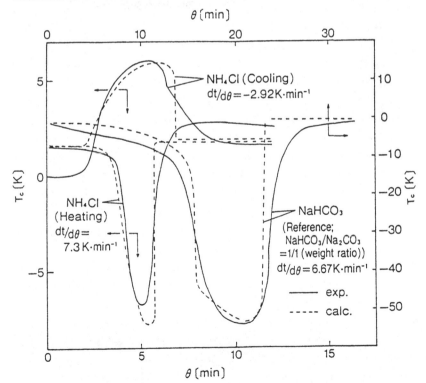

Masanobu Hasatani

Professor Hasatani is internationally known in the areas of combustion, fluidized beds, gas–solid reactions, and energy utilization. He currently is a professor of Chemical Engineering at Nagoya University.

Hitoki Matsuda

Professor Matsuda is internationally known for his work in thermal energy storage, chemical heat pumps and pipes, and energy utilization. He is currently associated with the Department of Chemical Engineering at Nagoya University.

EXPERIMENT 6

Temperature measurements in a transparent material: Application of holographic interferometry

Contributed by
AKIHIKO ITO

Principle

The measurement technique is based on (1) refractive-index changes in the material due to temperature variations, and (2) holographic recording to overcome the poor optical quality of the materials.

Object

The steady-state temperature distribution in a fused silica is measured as a demonstrative application of the technique. A brief quantitative evaluation of an interferogram is explained. Various errors in the measured temperature distribution caused by refraction and heat losses are discussed and estimated. The application to the measurement of 2D unsteady-state temperature distributions in a polymer and in a liquid are straightforward as demonstrated in Refs. 1–3.

Background

Let the sample prior to heating have length L in the direction of the laser beam (z direction, see Fig. 6.1) and a length of $L + l$ after heating. Given a two-dimensional heated sample with no variation of refractive index in the z direction and no light ray curvature, the difference in optical path length, ΔL^*, of the sample before and after heating is as follows:

$$\Delta L^* = L[n_{mT}(x,y) - n_{m\infty}] + l[n_{mT}(x,y) - n_{a\infty}] \qquad (6.1)$$

where n_m is the refractive index of the material, and n_a is the refractive index of the surrounding medium. Subscripts T and ∞ denote the temperature for heated material and the initial conditions (unheated), respectively. The first term on the right-hand side in the above equation is the change in optical path length due to the change in refractive index. The second term is the

33

length change of the sample due to its thermal expansion. They are expressed as:

$$L(n_{mT} - n_{m\infty}) = L\int_{T_\infty}^{T} \frac{dn_m}{dT} dT, \tag{6.2}$$

$$1 = \alpha L(T - T_\infty)$$

where dn_m/dT is the thermo-optic coefficient and α is the linear thermal expansion coefficient. The change in optical path length, ΔL^*, is related to the number of fringes by the relation

$$\Delta L^* = \frac{\lambda(2N - 1)}{2}, \qquad N = 1, 2, 3, \ldots \tag{6.3}$$

where N is the fringe number and λ is the wavelength of the light.

Substituting Eqs. (6.1) and (6.2) into Eq. 6.3, we obtain:

$$L\int_{T_\infty}^{T} \frac{dn_m}{dT} dT + \alpha L(T - T_\infty)\left[\int_{T_\infty}^{T} \frac{dn_m}{dT} dT + n_{m\infty} - n_{a\infty}\right]$$

$$= \frac{\lambda(2N - 1)}{2} \tag{6.4}$$

The first term in the bracket, $\int_{T_\infty}^{T} (dn_m/dT)dT$, is much smaller than $(n_m - n_a)$. If dn_m/dT is assumed to be constant, the relationship between the temperature T and the fringe number N can be approximately expressed as:

$$\Delta T = T - T_\infty = \frac{\lambda(2N - 1)}{2L[dn_m/dt + \alpha(n_{m\infty} - n_{a\infty})]} \tag{6.5}$$

The data on the refractive index versus temperature reported by Waxler and Cleek[4] were fitted to second degree polynomials. By differentiating this equation, we obtained the following equation:

Fig. 6.1. Schematic illustration of the sample and the coordinates for interferogram analysis and temperature and refractive-index distributions.

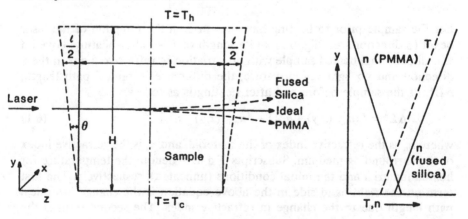

$$\frac{dn_m}{dT} \times 10^6 = 8.46 + 7.90 \times 10^{-3} \, \Delta T \, (^\circ C^{-1})$$

where $\Delta T \equiv T - 20\,^\circ C$. The linear thermal expansion coefficient α is $5.5 \times 10^{-7}\ (^\circ C^{-1})$. Now we can determine the temperature from Eq. (6.4) by counting the fringe number N.

Apparatus

A schematic illustration of the experimental apparatus is shown in Fig. 6.2.

A test sample is held between two aluminum blocks.

The upper aluminum block is heated by a well-regulated hot plate and the lower aluminum block is cooled by water flowing through it.

Both sides of the sample in the x direction are insulated by glass wool fiber.

50 μm diameter wire C-A thermocouples are used to measure the temperature of the upper and lower surfaces of the sample.

Commercially available fused silica (Dynasil), which is 3.2×3.8 cm^2 (width × height) with a 2.54 cm thickness (path length), is used as a test sample.

The optical setup for the holographic interferometer is shown schematically in Fig. 6.3.

The light source is a 5-mW He–Ne laser; the laser beam is divided into two

Fig. 6.2. Schematic illustration of experimental apparatus: TC = thermocouple.

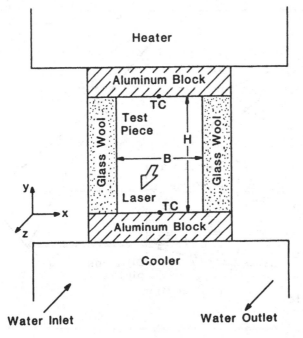

beams by a partially reflecting mirror. The object and reference waves inter-fere with each other at the holographic plate.

Procedure

A double-exposure method is used in this test

The first holographic exposure is made with the sample at ambient tem-perature.

After the sample is heated to and maintained at a steady state, a second holographic exposure is made.

The holograms are recorded on an Agfa-Gevaert 10E75 glass plate with an exposure time of 1/30s. These are developed for 7 min. in Kodak D-19.

Results

A typical result of an interferogram in fused silica is shown in Fig. 6.4(a). The temperature distribution in the *y* direction at the center of the sample width is also shown in Fig. 6.4(b). The squares in Fig. 6.4(b) represent the surface

Fig. 6.3. Optical system used for holographic interferometry: B.S. = beam splitter, B.E. = beam expander, C.M. = concave mirror, F = filter, M = mirror, S = shutter.

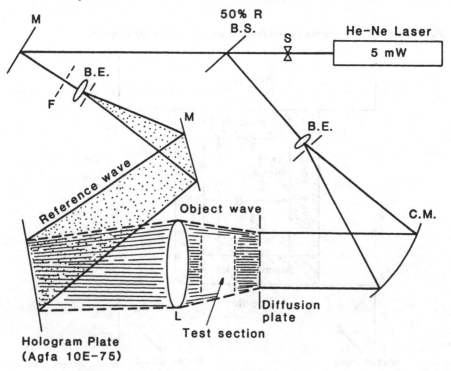

Fig. 6.4. (a) Typical interferogram and (b) temperature distribution in fused silica.

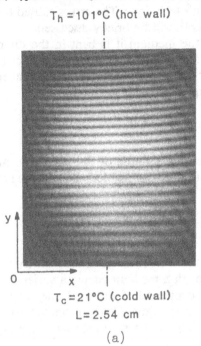

$T_h = 101°C$ (hot wall)

y

0   x

$T_c = 21°C$ (cold wall)

L = 2.54 cm

(a)

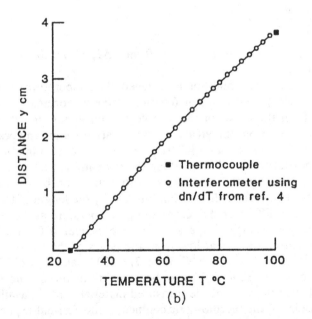

(b)

temperature of the sample measured by the thermocouples. There are several potential sources of error in this technique. The errors caused by refraction and heat losses from the sample will be briefly discussed.

As a light ray passes through the medium it is bent in the direction of increasing index (see Fig. 6.1). If the refractive index dn_m/dT can be approximated as a linear function of the y coordinate, the differential equation describing the path ray can be expressed as

$$\frac{d^2y}{dz^2} = \frac{1}{n_m}\frac{dn_m}{dy} \simeq a + by \tag{6.6}$$

where a and b are constants. When the solution of Eq. (6.6) is expanded in a Taylor series and higher-order terms are disregarded, the following equation is obtained:

$$\delta = \frac{aL^2}{2}\left(1 + \frac{bL^2}{12}\right) \tag{6.7}$$

where δ is the difference between the y coordinates of the light entrance and exit points. As the ray passes through a medium having a variation in temperature caused by the deflection expressed of Eq. (6.7), the fringe shift is proportional to the path-averaged temperature. Then the temperature error, ΔT_ε, caused by the deflection as in Eq. (6.7) can be approximately expressed as follows:

$$\Delta T_\varepsilon = \frac{aL^2}{6}\frac{dT}{dy}\left(1 + \frac{bL^2}{20}\right) \tag{6.8}$$

In this test the maximum errors are $\delta = 4.2 \times 10$ cm^{-4}, $\Delta T_\varepsilon = 1.3 \times 10^{-3}$ °C. These errors are negligible.

Even though a steady-state condition is achieved, the temperature distribution shown in Fig. 6.4(b) is not a linear function of the y coordinate. This is due to heat loss from the edges of the sample by convection and thermal radiation. This heat loss from the edges results in variations in temperature along the ray path. For this reason, the fringe number is determined by an integral of the temperature distribution along the ray path. An average temperature for this nonisothermal temperature distribution can be calculated from the integrated temperature distribution divided by the length of the ray path. The temperature difference ΔT_ξ between the temperature at the center of the sample and the average temperature can be estimated from a heat balance equation that includes heat loss from the edges of the sample. It is found that the maximum ΔT_ξ is 0.15 °C when $T_h = 100$ °C and $T_c = 20$ °C.

The technique does not require high optical quality material. The technique requires: (1) transparency of the measured material, and (2) availability of the relationship for the thermo-optic coefficient and thermal-expansion coefficient of the material with temperature. If these requirements are satisfied, the temperature distribution of the material can be measured with a high spatial resolution.

Suggested headings

Sample length $L =$ _____; Wavelength $\lambda =$ _____

Refractive index of material at T_∞ $n_{m,\infty} =$ _____

Refractive index of the surrounding medium $n_{a\infty} =$ _____

Ambient temperature $T_\infty =$ _____

Distance		Fringe number	Temperature	
x	y	N	ΔT $(\equiv T - T_\infty)$	T

References

1. Ito, A., and Kashiwagi, T., "Temperature measurements in PMMA during downward flame spread using holographic interferometry," *Proc. Twenty-First Symposium (International) on Combustion*, Combustion Institute, Pittsburgh, PA, 65–74, 1986.
2. Ito, A., and Kashiwagi, T., "Measurement technique for determining the temperature distribution in a transparent solid using holographic interferometry," *Applied Optics* 26, 5 (1987): 954–8.
3. Ito, A.; Masuda, D., and Saito, K., "A study of flame spread over alcohols using holographic interferometry," *Combustion and Flame* 83 (1991): 375–89.
4. Waxler, R. M., and Cleek, G. W., "The effect of temperature and pressure on the refractive index of some oxide glasses," *J. Res. Natl. Bur. Stand. Sec. A* 77 (1973): 755–63.

Akihito Ito

Professor Ito is an associate professor in the Department of System Production Engineering, Oita University. He received his D.Eng. in mechanical engineering from the Tokyo Institute of Technology in 1979. He worked as a research associate in Kyushu University until 1982. He was involved in combustion and fire research studies at the National Bureau of Standards in 1985 and the University of Kentucky in 1991.

PART I.2
Convection

Fig. I.2. Convection in a rotating annulus with negative radial temperature gradient. The core is 15 °F warmer than the rim. Rotation is 2.5 radians per second counterclockwise. Aluminum powder shows four symmetric waves produced by baroclinic instability. (Courtesy of E. L. Koschmieder, *J. Fluid Mech*. 51 (1972): 637.)

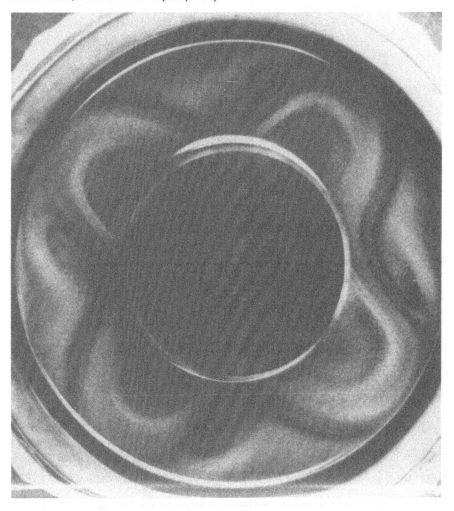

EXPERIMENT 7
A forced-convection heat-transfer experiment

Contributed by
W. H. GIEDT

Principle

The action of the driving fan or compressor introduces significant turbulence in a wind-tunnel test stream. This is illustrated by finding that the local heat transfer along a flat plate parallel to a wind-tunnel test stream is characteristic of turbulent flow even though $Re_x < 1 \times 10^5$.

Object

The object of this experiment is to calculate the heat-transfer coefficient over a flat plate of zero angle-of-attack utilizing a transient technique.

Background

The theoretical background for convective heat transfer over a flat plate is well established and can be found in any basic heat-transfer text. The standard technique is to assume the velocity and temperature profiles in the boundary layer over a flat plate for either laminar or turbulent flow. For example, if one assumes a cubic (four-term polynomial) or the velocity distribution, the laminar-boundary-layer thickness δ is determined as

$$\delta = 4.64x/\sqrt{Re_x} \tag{7.1}$$

Using a cubic (four-term polynomial) for the temperature distribution in a laminar flow, the thermal-boundary-layer thickness δ_t is then found to be

$$\delta_t = 0.976\delta \, Pr^{-1/3} \tag{7.2}$$

The rate of heat flow by convection from the plate per unit area is, for a cubic approximation of the temperature profile,

$$q/A = -k \left.\frac{\partial T}{\partial y}\right|_{y=0} = \frac{3}{2}\frac{k}{\delta_t}\,\Delta T \tag{7.3}$$

43

The local Nusselt number Nu_x is then

$$Nu_x = \frac{h_x x}{k} = \frac{q_x x}{kA \, \Delta T} = 0.33 \, Re_x^{1/2} \, Pr^{1/3} \tag{7.4}$$

This is almost identical to exact solution for laminar flow over a flat plate,

$$Nu_x = 0.332 \, Re_x^{1/2} \, Pr^{1/3} \tag{7.5}$$

For fully turbulent flow over a flat plate, in the local Reynolds number range of $5 \times 10^5 < Re_x \leq 10^7$, one finds

$$Nu_x = 0.0296 \, Re_x^{0.8} \, Pr^{1/3} \tag{7.6}$$

for $0.6 < Pr < 60$.

During cooling of a thin flat plate suddenly exposed in an air stream, the local heat-transfer coefficient h_x can be calculated from the energy rate balance (assuming conduction along the plate is negligible)

$$-\rho \frac{\bar{t} c}{2} \frac{dT_x}{dt} = h_x \, (T_x - T_a) \tag{7.7}$$

where $\bar{t}$ is the thickness of the plate, ρ, c are the plate's density and specific heat capacity, respectively, and dT_x/dt represents the instantaneous local rate of temperature change.

Apparatus

Stainless-steel flat plate (selected because its thermal conductivity is low)
Stainless-steel rectangular duct
Oven
Wind tunnel
Four copper-constantan thermcouples

Procedure

Install the four thermocouples along the midplane of the plate according to the geometry shown in Fig. 7.1 (no. 36 wire size suggested.)
Mount the stainless-steel plate vertically in a rectangular duct (also of stainless steel).
Heat the plate and duct in an oven to approximately 200 °F.
Rapidly move the heated duct and plate from the oven and attach the assembly to the exit of a small wind tunnel.
Record the thermocouple voltages (referenced to ambient air temperature) versus time. If there is good electrical contact between the thermocouples and the plate, electrical currents will flow from hot to cooler

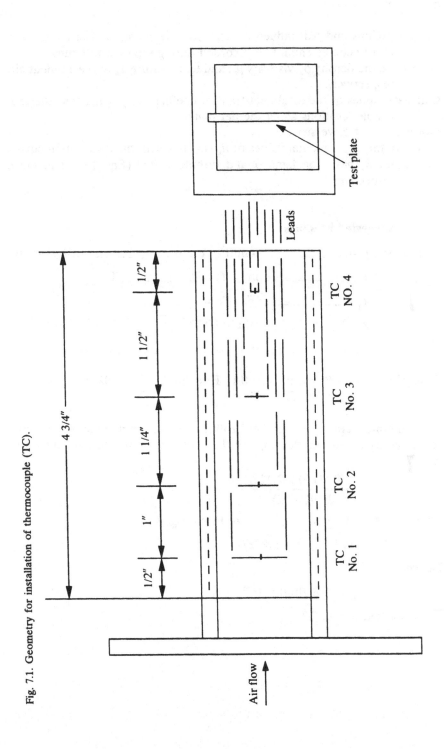

Fig. 7.1. Geometry for installation of thermocouple (TC).

regions and will influence thermocouple readings. Therefore, each thermocouple should be recorded during separate test runs.

Determine the density ρ_a, viscosity μ_a, and temperature T_a of the ambient air, respectively.

Calculate values of the local heat-transfer coefficient h_x at the four thermocouple locations along the flat plate.

Plot a curve of b_x versus x.

Compare the experimental values of h_x versus x with the distribution curves for h_x based on laminar and turbulent flow (Eqs. (7.5) and (7.6), respectively).

Suggested headings

Constants: $\rho =$ _____ lbm/ft³; $c =$ _____ Btu/1bm°F

$\bar{t} =$ _____ in.; $T_a =$ _____ °F

$T_p =$ _____ °F; $Pr =$ _____

$k =$ _____ Btu/hr · ft°F

$U_\infty =$ _____ ft /s; $\Delta p =$ _____ mm H₂O

Example: $\rho = 488$ lbm/ft³, $c = 0.11$ Btu/1bm °F, $\bar{t} = 0.102$ in., $T_a = 72$ °F, $T_p = 191$ °F.

For thermocouples 1 and 2, Fig. 7.2 illustrates the temperature drop versus real time. By measuring the slope of the curve, one can determine dT_x/dt in Eq. (7.7).

$$h_x = \frac{488(0.11)\,(0.102/12)T'}{2(191 - 72)}$$

where TC #1 is located at $x = 0.5$ in. From Fig. 7.2,

$$T' = dT/dt|_{x=0.5\ in.} = 4.898\ °F/s.$$

Hence

$$h_x|_{x=0.5\ in.} = 33.8\ \text{Btu/hr ft}^2\ °F$$

and is plotted in Fig. 7.3.

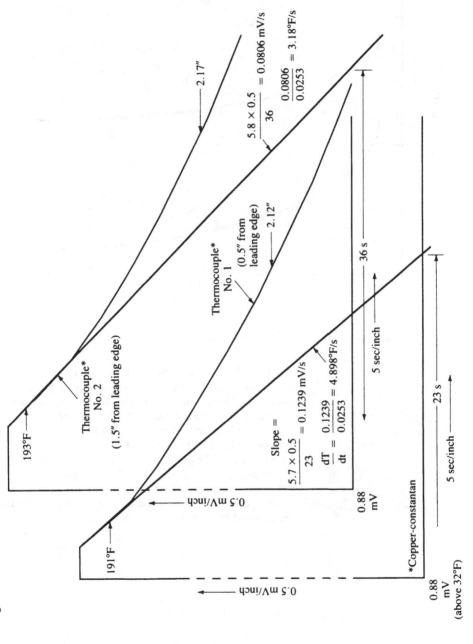

Fig. 7.2. 0.102-inch-thick stainless-steel plate cooled in airstream at 72 °F.

Fig. 7.3. Heat-transfer coefficient h_x versus position.

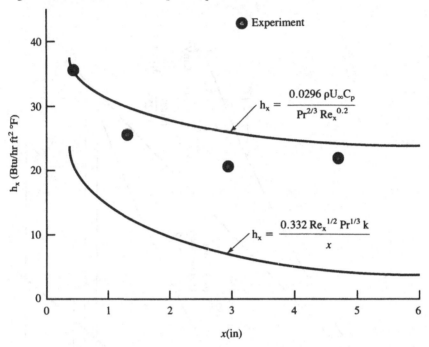

● Experiment

$$h_x = \frac{0.0296\ \rho U_\infty C_p}{Pr^{2/3}\ Re_x^{0.2}}$$

$$h_x = \frac{0.332\ Re_x^{1/2}\ Pr^{1/3}\ k}{x}$$

x(in)

h_x (Btu/hr ft^2 °F)

Warren H. Giedt

Warren H. Giedt is Professor Emeritus in the Department of Mechanical, Aeronautical, and Materials Engineering of the University of California at Davis and a consultant to the Sandia and Lawrence Livermore National Laboratories. He was editor of the ASME *Journal of Heat Transfer* from 1967 to 1972. His publications include two textbooks, *Principles of Engineering Heat Transfer* and *Thermophysics*, and more than 125 technical papers and reports. His professional contributions have been recognized by numerous awards.

EXPERIMENT 8
Reynolds analogy for mass transfer

Contributed by
L. C. BURMEISTER

Principle

The convective heat- and mass-transfer coefficients for evaporating water are
related by the Reynolds analogy.

Object

The object of this experiment is to determine the ratio of the convective heat-
and mass-transfer coefficients by measuring the rates of cooling of two cups
of water simultaneously exposed to the same air stream, one cup with evaporation and one without evaporation.

Background

Consider a cup filled with water. It is insulated on the bottom and sides,
convectively losing heat and mass to the air only from the open top. The
initial water temperature T_i is greater than the ambient air temperature T_o.

The variation of water temperature T with time t elapsed since cooling began
is described by the conservation of energy principle applied to the cup. Thus,

> Rate of Energy Storage
> = Rate of Energy Gain by Convective Heat Transfer
>
> +
>
> Rate of Energy Gain by Convective Mass Transfer

In terms of the water mass m, specific heat C_p, heat of vaporization h_{fg}, top
surface area A, convective heat-transfer coefficient h, convective mass-
transfer coefficient h_D, mass fraction ω_1 of the water vapor, and air density ρ
this relationship is

$$mC_p dT/dt = -hA(T - T_o) - h_{fg}\rho h_D A(\omega_{1w} - \omega_{1o})/(1 - \omega_{1w})$$

subject to the initial condition $T(t = 0) = T_i$. Here the water vapor on the air
side of the liquid-water–air interface is taken to be species 1 of the binary

air–water-vapor mixture, the subscript w refers to conditions on the air side of the liquid-water–air interface, and the subscript o refers to ambient conditions. Rearrangement, assuming $T - T_o \simeq T_i - T_o$, gives

$$\tau d\theta/dt = -\theta, \quad \theta(t = 0) = 1 \tag{8.1}$$

where $\theta = (T - T_o)/(T_i - T_o)$ and the time constant τ is

$$\tau = \tau_h/\{1 + \rho h_{fg}(h_D/h)(\omega_{1w} - \omega_{1o})/[(1 - \omega_{1w})(T_i - T_o)]\} \tag{8.2}$$

Here $\tau_h = mC_p/hA$ is the time constant for the cooling process without evaporation, as would be the case if the liquid water were covered with a thin film of oil.

The mass fraction ω_1 of the water vapor is related to the partial pressure p_1 of the water vapor by

$$\omega_1 = p_1(M_1/M_2)/[p - p_1(1 - M_1/M_2)] \tag{8.3}$$

where M_1 is the molecular weight of water, M_2 is the molecular weight of air, and p is the total atmospheric pressure. At the liquid-water–air interface, p_1 equals the saturation pressure $p_v(T)$ of water at the water temperature, enabling ω_{1w} to be evaluated. In the ambient air, p_1 is equal to $\phi p_v(T_o)$ where ϕ is the relative humidity, enabling ω_{1w} to be evaluated. Provided that the time constant τ is constant, the solution to Eq. (8.1) is $\ln(\theta) = -t/\tau$.

Also, the Reynolds analogy provides the relationship

$$Nu/RePr^{1/3} = Sh/ReSc^{1/3} = C_f/2 \tag{8.4}$$

where the Nusselt number Nu is related to the air thermal conductivity k and a characteristic length L by

$$Nu = hL/k \tag{8.5}$$

the Sherwood number Sh is related to the mass diffusivity D_{12} of water vapor through air by

$$Sh = h_D L/D_{12} \tag{8.6}$$

the Prandtl number Pr is related to the diffusivity for momentum (kinematic viscosity) v and thermal diffusivity α of air by

$$Pr = v/\alpha \tag{8.7}$$

and the Schmidt number Sc is given by

$$Sc = D_{12}/\alpha \tag{8.8}$$

The Reynolds number Re is related to a characteristic fluid velocity V by

$$Re = VL/v \tag{8.9}$$

and the friction coefficient C_f is related to shear stress σ_w at the liquid-water–air interface by

$$\sigma_w = C_f \rho V^2 / 2 \tag{8.10}$$

although neither Re nor C_f is needed in the present investigation. From the Reynolds analogy, it is seen that

$$h_D/h = (Sc/Pr)^{1/3} D_{12}/k \tag{8.11}$$

Apparatus

Two paper cups
Milliliter measuring beaker
Two thermometers
Sling psychrometer
Light oil
Expanded fiberglass insulation

Procedure

Insulate the bottom and side of each of the two cups with the expanded fiberglass insulation.

Determine the relative humidity and dry-bulb temperature of the ambient air with the sling psychrometer.

Fill both cups from the beaker with equal known volumes of water about 10–20 °C warmer than the ambient air. Add enough light oil to one cup to form a *thin* oil film on the water surface.

Insert a thermometer in each cup.

Record the temperature of the water in each cup at frequent intervals. Gently stir the water in each cup between measurements.

On semilogarithmic coordinates, plot the dimensionless water temperature θ as the ordinate versus time t elapsed since the start of the cooling as the abscissa.

Determine the numerical value of τ from the plot described in the prior step for the water temperatures of the cup without the oil film. Note from Eq. (8.1) that since $\ln(\theta) = -t/\tau$ if the time constant τ is constant, a straight line passing through the $\theta(t=0) = 1$ point and through nearby data points intercepts the $\theta = 1/e$ level at a value of t equal to the time constant. Evaluate the time constant τ_h in a similar manner from the water temperatures of the cup with an oil film.

Compare the value of the ratio h_D/h of the convective mass-transfer coefficient to the convective heat-transfer coefficient evaluated by means of transient temperature measurements and Eq. (8.2) with that evaluated by means of the Reynolds analogy, Eq. (8.11).

Questions

1. Is the time constant actually constant for either of the two cups?
2. Is the thermal resistance provided by the fiberglass insulation large enough to ensure that at least 95 percent of the heat transfer occurs across the liquid-water–air interface?
3. Would a different value of the ratio h_D/h be obtained if the initial temperature of the water were increased to 30–40 °C above the ambient temperature?
4. Is it possible to determine the numerical value of the convective heat-transfer coefficient h from the data for the cup with an oil film? If so, how could this result be used to determine the numerical value of the convective mass-transfer coefficient h_D?
5. How important is the $1 - \omega_{1w}$ term, representing the effect of water vapor "blowing" into the air at the liquid-water–air interface, in Eq. (8.2)?

Suggested headings

Constants:

$A =$ _____ m^2; $m =$ _____ kg; $T_o =$ _____ °C; $\phi =$ _____

$D_{12} =$ _____ m^2/s; $k =$ _____ W/m K°; $Sc =$ _____ ; $Pr =$ _____

Data:

| | With oil film | Without oil film |
$t(s)$	$T(°C)$	$T(°C)$

Parameters derived from data:

$\tau =$ _____ s; $\tau_h =$ _____ s

Reference

Burmeister, L. C., 1983, *Convective Heat Transfer*, New York, Wiley-Interscience, pp. 173, 328, 331.

Louis C. Burmeister

Professor of mechanical engineering at the University of Kansas, Dr. Burmeister's primary interests are in the heat-transfer area. He has written a leading textbook on convective heat transfer and papers on valve technology, film boiling, vibration of compliant tanks, clothes washing machine design, solar energy, rubber-band heat engines, Monte Carlo methods for heat conduction, cogeneration, power-factor correction with capacitors, natural convection in porous media, and flow in plastic injection molding machines.

EXPERIMENT 9
Natural-convection melting of a slab of ice

Contributed by
ADRIAN BEJAN *and* ZONGQIN ZHANG

Principle

The heat transfer across the air boundary layer that descends along a vertical ice slab causes melting at the surface.

Objective

The effect of heat transfer by boundary-layer natural convection over a vertical wall can be visualized and measured by experimenting with thin slabs of ice suspended vertically in still air. The uneven distribution of heat flux is demonstrated by the uneven thinning of the ice slab. The instantaneous flow rate of meltwater collected under the dripping ice is a measure of the overall heat transfer rate from the ambient to the isothermal surfaces of the slab. An additional objective of this experiment is to show that laboratory apparatuses can be built quite inexpensively, often by using kitchen utensils. This experiment teaches a group of students to critically evaluate each others' data, and to pool all their findings into a comprehensive report that may have engineering significance.

Apparatus

Baking pan
Refrigerator
String
Cardboard box
Sheet-metal tray
Thermometer
Graduated beaker
Clock

The heart of the apparatus is a vertical slab of ice, which is suspended by means of a string in still air. The manufacture of the ice slab and its suspension

54

and the maintenance of a nearly motionless and isothermal ambient are the critical aspects of the apparatus construction.

An inexpensive way of producing ice slabs of one or more sizes is to use a flat-bottom baking pan (or cookie sheet) placed horizontally in the freezer of a household refrigerator. Desirable ice slab qualities are

(i) a temperature close to the melting point 0 °C, in other words, a minimum degree of solid subcooling, and
(ii) a minimum amount of trapped air bubbles and other defects (cracks, bulges) in the free surface.

The first feature is enhanced by using a freezer the temperature setting of which reads relatively "warm." The second feature is more difficult to attain, as it requires the sequential freezing of thin layers of water no deeper than approximately 0.5 cm.

Figure 9.1 shows the main steps in the production of one ice slab. In the first step, the slab is frozen (built up) to its half-thickness $L/2$. At this point, the suspension string (total length ~1 m) is looped and placed flat over the free surface of the frozen slab. The two ends of the string hang over the side of the tray, by intersecting that edge of the ice slab that in the actual experiment will face upward.

In the second step, the remainder of the slab thickness L is built up through the freezing of additional water layers. For a slab of final thickness $L = 1$ cm,

Fig. 9.1. Construction of the experimental apparatus.

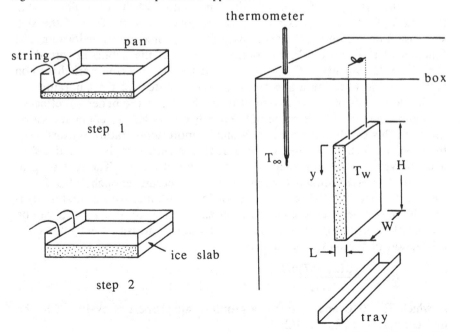

it is sufficient to freeze a single 0.5 cm-deep water layer in the first step, followed by another 0.5 cm-deep layer in the second step.

The baking pan dictates the large dimensions of the ice slab, namely, the height H and the width W. The very slow process of natural-convection melting – the actual experiment – begins by suspending the ice slab in still air. An inexpensive type of enclosure that prevents the forced convection of air is a cardboard box. The suspension of the ice slab inside this box is achieved by passing the two ends of the string through two holes in the ceiling of the box, and tying them into a knot on the outside.

Positioned under the ice slab is a sheet metal tray (trough), which catches the water droplets that fall. This tray is inclined relative to the horizontal, so that the collected liquid passes through an opening in the side of the cardboard box, and falls into a graduated beaker, or measuring glass column.

Procedure

The quantity that is measured during this experiment is the meltwater flow rate $\dot{m}$. This measurement is obtained by monitoring the rise of the water level in the graduated beaker. It is a relatively simple measurement, as the melting rate is expectedly slow. For example, if the slab dimensions H and W are of the order of 30 cm, then the volumetric flow rate of meltwater is of the order of 200 cm^3/h. It is a good idea to wait at least 30 minutes before measuring $\dot{m}$, in order to allow the transient conduction inside the slab to run its course, and bring the entire slab volume to the melting point.

The other quantity of interest is the temperature of the air maintained in the box T_∞ (°C). This air is not sealed off completely, because of the slab suspension device and the opening used for the collection of meltwater. The function of this "leaky" box is simply to prevent the action of forced air currents in the vicinity of the vertical ice surface. Such currents are driven on and off by the ventilation and air-conditioning system of the laboratory.

The slow leaks of air into and out of the box have the beneficial effect of regulating the bulk temperature of the nearly motionless air contained inside the box. This can be measured with one or more temperature sensors (e.g., thermometers). It turns out that one sensor is sufficient, because the slow leaks prevent the vertical thermal stratification of the air. The reading provided by a mercury-in-glass thermometer is accurate enough (±0.3 °C), in view of the fact that T_∞ is of the order of 15 °C. In other words, the temperature difference between the air reservoir and the ice surface, $T_\infty - T_w$, can be estimated to within ±2 percent. This temperature measurement allows the calculation of the Rayleigh number

$$Ra_H = \frac{g\beta(T_\infty - T_w)H^3}{\alpha v} \tag{9.1}$$

in which $T_w = 0$ °C and $(\beta/\alpha v)$ is a group of air properties evaluated at the film temperature $(T_\infty + T_w)/2$.

Explanation

The instructional value of this very simple experiment lies in the visual observations afforded by the melting slab (Fig. 9.2), and in the effort of predicting (anticipating theoretically) the melting rate $\dot{m}$.

Two long windows cut into the side of the box, and covered with plastic wrapping material (of the kind used in the kitchen), allow the experimentalist to look in the direction parallel to the large surfaces of area $H \times W$. As the time passes, it is observed that the slab melts *unevenly*: Near its upper edge, the slab becomes considerably narrower than over the remainder of its height.

The previous observation finds an explanation in the argument that the change of phase at the slab surface is driven mainly by the local heat flux from the T_∞ air to the T_w surface, across a boundary layer of cold air that descends along the wet ice surface,

$$q''(y) = \rho_w h_{sf} \frac{d}{dt}\left(\frac{L(y)}{2}\right) \tag{9.2}$$

In this equation, $\rho_w \cong 1$ g/cm^3 is the density of water at 0 °C, and $h_{sf} = 333.4$ J/g is the latent heat of melting of ice. The uneven shape of the instantaneous slab half-thickness $L(y)/2$ is the time-integrated effect of the local free-convection heat flux $q''(y)$. In the laminar regime, q'' decreases as $y^{-1/4}$ in the downstream direction (vertically downward in Figs. 9.1 and 9.2).

The meltwater flow rate $\dot{m}$ can be predicted by noting its geometric definition,

$$\dot{m} = \rho_w HW\left(\overline{\frac{dL}{dt}}\right) \tag{9.3}$$

where $(\overline{dL/dt})$ is the rate of slab thinning, averaged over the height H. Combined, Eqs. (9.3) and (9.2) yield

$$\dot{m} = 2W\frac{k}{h_{sf}}(T_\infty - T_w)(\overline{Nu_H}) \tag{9.4}$$

where $\overline{Nu_H}$ is the overall Nusselt number based on the height-averaged heat flux $\overline{q''}$:

$$\overline{Nu}_H = \frac{\bar{h}H}{k} = \frac{\overline{q''}H}{(T_\infty - T_w)k} \tag{9.5}$$

In Eqs. (9.4), (9.5), k is the thermal conductivity of air evaluated at the film temperature. The overall Nusselt number can be calculated using one of the textbook formulas $\overline{Nu}_H(Ra_H, Pr)$ for laminar boundary layer natural convection over a vertical plane wall.

It is likely that the experimentally measured melting rate $\dot{m}$ will be somewhat greater than the value anticipated using Eq. (9.4). This discrepancy should be commented on. It may be caused, for example, by the effect of

Fig. 9.2. The gradual thinning of an ice slab, during melting by natural convection ($H = 38.3$ cm, $W = 25.5$ cm, $T_\infty = 18\ ^\circ$C, $\dot{m} = 228$ g/h; from left to right, the elapsed times are 25, 70, 115, and 160 minutes).

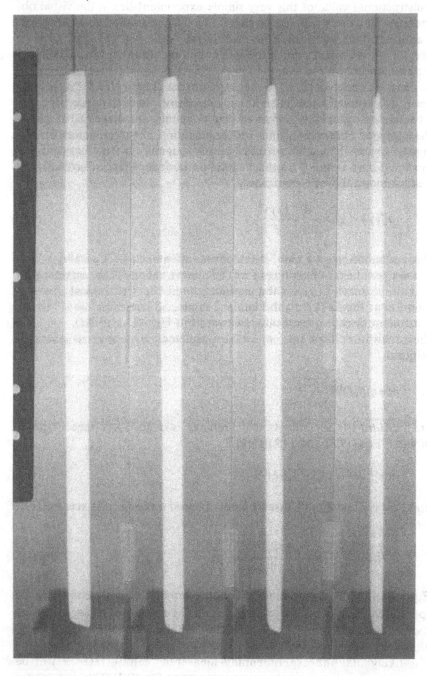

melting along the four narrow surfaces of the slab, and by the radiation heat transfer between the slab and the room-temperature wall of the cardboard box. The radiation effect can be estimated in an order of magnitude sense, by using the two-surface gray enclosure theory.

Of particular interest in ice storage applications and meteorology is the relationship between the melting rate and the bulk temperature of the surrounding air. The present experiment reveals this information as a relationship between $\dot{m}$ and Ra_H. It is considerably easier to vary the Rayleigh number by changing the vertical dimension of the ice slab, as opposed to changing the imposed temperature difference $(T_\infty - T_w)$.

The $\dot{m}$ (Ra_H) curve can be pinpointed by several students (or groups of students) who monitor the melting of ice slabs with several different slab heights H. Worth noting is that a single baking pan can produce ice slabs for two different Ra_H cases, depending on which side of the large rectangular surface is oriented vertically during the experiment. By using ice slabs with heights H in the range 0.25–1.25 m, it is possible to vary Ra_H by two orders of magnitude, in the range 10^7–10^9.

It is recommended that at the end of the experimental session each student reports his or her experimental point $(\dot{m}, Ra_H)$ on one line in the following table. After the session, and before the handing in of the individual experimental report, the student should analyze the completed table. One question to pursue is whether the $\dot{m}(Ra_H)$ curve is anticipated adequately by the theory of Eqs. (9.2)–(9.5). Furthermore, by plotting the $(\dot{m}, Ra_H)$ data in dimensionless logarithmic coordinates (M, Ra_H), where

$$ M = \frac{\dot{m} h_{sf}}{W k (T_\infty - T_w)} \tag{9.6} $$

it is possible to tell if M is proportional to $Ra_H^{1/4}$, in accordance with laminar boundary layer theory.

This final exercise gives each student a global perspective on the success of the experimental session, for example, on who agrees with whom, and why. It also teaches that disagreements between two or more experimentalists are very much part of the game, and, if these disagreements cannot be reasoned away, it is time to devise a better theory!

Suggested headings

Name	H [cm]	W [cm]	T_∞ [°C]	$\dot{m}$ [g/h]	M	Ra_H

Adrian Bejan

Adrian Bejan is J. A. Jones Professor of Mechanical Engineering at Duke University. He was educated at the Massachusetts Institute of Technology (B.S., M.S., Ph.D. degrees), doing graduate research in superconductivity and cryogenics. In 1976–8 he was a postdoctoral Miller Fellow at the University of California, Berkeley, where he began his fundamental research in heat transfer, thermodynamics, and fluid mechanics. He is the author of 140 technical papers and three graduate textbooks, *Entropy Generation through Heat and Fluid Flow, Convection Heat Transfer,* and *Advanced Engineering Thermodynamics.*

EXPERIMENT 10
Forced-convection heat loss from 3D solids

Contributed by
RICHARD I. LOEHRKE

Principle

Newton's law of cooling: The rate of heat loss from a warm body cooled by a fluid stream is proportional to the difference in temperature between the surface of the body and the fluid far away from the body. See Ref. 1.

Object

The object of this experiment is to determine the average heat transfer co-efficient at the surface of a sphere and other compact 3D solids from measurements of the solid temperature history in a transient thermal test.

Background

The rate of heat loss from a surface to a fluid is controlled by diffusion processes in the fluid, modified by fluid motion. The rate should scale directly with overall temperature difference if the fluid motion and thermal properties are temperature independent and provided that the thermal boundary conditions remain similar as the temperature difference changes.

In this experiment, a warm solid body is placed in a constant-velocity air stream and allowed to cool. The body is made of copper so that internal temperature gradients are small and a single temperature T can be used to characterize the solid. The upstream fluid temperature T_f is constant. A heat balance on the solid, neglecting radiation losses, gives the relation

$$h \equiv \frac{q}{A(T - T_f)} = - \frac{\rho V c_p}{A(T - T_f)} \frac{dT}{dt} \tag{10.1}$$

where q is the rate of heat loss from the body which has surface area A and volume V, ρ is the density and c_p the specific heat of the body, t is time, and h is the convective heat transfer coefficient. If the properties of the solid body are known and if the fluid temperature and solid temperature histories are

61

measured, then the instantaneous h can be determined. According to Newton's law of cooling, h should not change with time.

The assumption of negligible internal temperature gradients is valid provided that the Biot number $hD/k_s \ll 1$, where D is a characteristic length of the solid and k_s is the solid conductivity. This can be converted into a rough estimate for the upper limit of Reynolds number Re permissible for this experiment as

$$\frac{hD}{k_s} = \left(\frac{hD}{k_f}\right)\left(\frac{k_f}{k_s}\right) \approx 0.5 \, (Re)^{0.5} \left(\frac{k_f}{k_s}\right) \ll 1 \tag{10.2}$$

where k_f is the fluid conductivity and where an approximate relation between the Nusselt number $Nu = hD/k_f$ and the Reynolds number for a sphere was used. For copper and air this inequality becomes $Re \ll 8.7 \times 10^8$.

Natural convection and radiation will become important at low Reynolds number. Then the effective heat transfer coefficient determined from Eq. (10.1) will no longer be independent of temperature. The lower limit on Reynolds number for Newtonian cooling depends on the size of the body; the smaller the body the lower the limit.

Apparatus

Air jet
Solid copper sphere, cube, and short cylinder (one each)
Thermocouple wire
Thermocouple indicator
Pitot tube
Manometer
Clock with second hand
Propane torch

Drill a shallow hole in each solid. Insert and solder a thermocouple in each hole.

The air jet may be provided by a small squirrel-cage fan blowing through a short section of duct as shown in Fig. 10.1. Screens stretched across the duct help to form a uniform-velocity profile at the exit. Use a 4-in. or larger jet with a 1-in. sphere.

Procedure

Set a constant jet velocity. Measure and record the velocity.
Measure the air temperature.
Heat one of the solids to approximately 150 °C with the torch.
Suspend the solid, by the thermocouple wire, in the air jet.
Record the temperature history between 150 °C and 50 °C.

Plot the measured temperature and the calculated heat transfer coefficient versus time. Data reduction is less tedious and the accuracy of the determination of an instantaneous h can be improved if an A/D data acquisition system is used. Then the derivative in Eq. (10.1) can be approximated with finite differences of closely spaced data as in Fig. 10.2.

Repeat the procedure at different velocities and with different solid bodies.

Questions

1. Does the measured heat transfer coefficient vary with body temperature at a constant air velocity? Compare natural convection (zero jet velocity) and forced convection.

Fig. 10.1. Squirrel-cage fan rigged to provide an air jet.

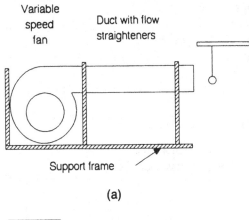

(a)

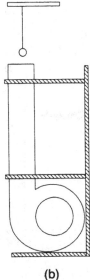

(b)

2. How do the results for the sphere compare with correlations available in the literature?
3. Does the presence of the thermocouple influence the rate of heat loss? Compare the results for the sphere obtained with the thermocouple normal to the air stream (horizontal jet) with those obtained with the thermocouple downstream from the sphere (vertical jet). See Fig. 10.1.
4. How does h vary with orientation for the asymmetric solids?
5. Can the results for all of the solids be described by a single correlation of Nusselt number with Reynolds number?
6. What is the effect on the heat transfer coefficient if the uniform velocity is disturbed by placing another, unheated body just upstream from the solid in question? What if the body upstream is also heated?

Suggested headings

Constants: Body – shape, V, A, ρ, c_p
              Air – U, T, p

Independent variable: t

Fig. 10.2. Data from a one-inch sphere in 20 °C air.

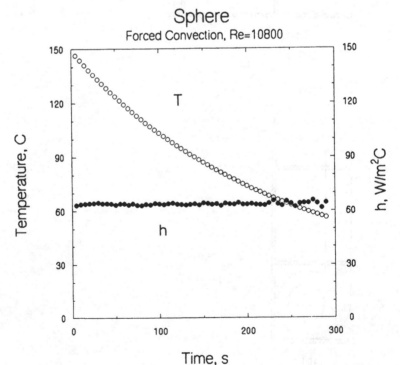

Measured variable: T

Calculated: h, Nu, Re

Reference

1. Bergles, A. E., "Enhancement of convective heat transfer: Newton's legacy pursued," *History of Heat Transfer*, eds. E. T. Layton, Jr. and J. H. Lienhard, ASME, New York, 1988, pp. 53–64.

Richard I. Loehrke

Richard I. Loehrke is a professor of mechanical engineering at Colorado State University. He served as a project engineer with the General Electric Co. in Evendale, OH, and as research engineer with Sundstrand Aviation in Denver, CO. He obtained a Ph.D. from the Illinois Institute of Technology. Professor Loehrke joined Colorado State University in 1971 where he has taught and conducted research in the areas of thermodynamics, heat transfer, and fluid dynamics.

EXPERIMENT 11
Forced and free convective heat transfer of a laminar flow in a horizontal heated pipe

Contributed by
YASUO MORI

Principle

The forced convective heat transfer of a laminar flow in a heated horizontal circular straight pipe is enhanced by buoyancy.

Object

The experiment demonstrates the importance of simultaneous consideration of flow and temperature fields in forced convective heat transfer. For an air flow in a circular horizontal straight pipe, when the wall temperature is over 1 °C higher than that of the flow, the measured flow profile is much different from Poiseuille due to the secondary flow by buoyancy, and the temperature profile is seen not to be axially symmetric. These results peculiar to the combined heat transfer are understood from examination of the profiles of velocity and temperature and through flow visualization.

Apparatus

Air blower
Orifice flow meter
Circular straight pipe of bare and heated sections
Pitot tube and manometer
Thermocouple probe and millivoltmeter
Smoke generator
Traversing device for yaw meter and thermocouple
Electric power course for heating the pipe

A schematic view of the experimental apparatus suitable for a small group demonstration is shown in Fig. 11.1. Air is used as the working substance. The main part of the apparatus consists of an air blower, a straight horizontal circular copper pipe of 30-mm inner diameter, 1-mm thickness, and 7-m total

Fig. 11.1. Schematic view of experimental apparatus.

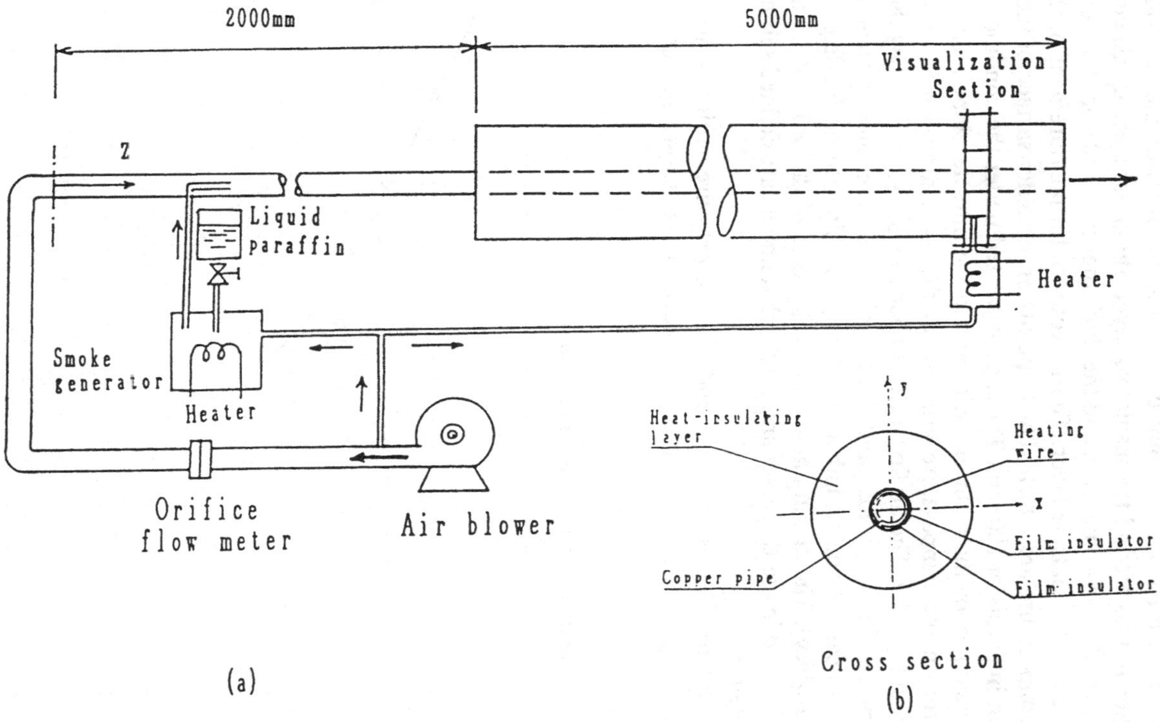

length. The upstream part of 2-m length is a bare pipe to provide for a fully developed laminar-flow region, while the remaining 5-m part is the heated part for measurement of heat transfer. The pipe of the heated portion has the four layers in piles around it, as shown in Fig. 11.1b, which are the thin electric-insulating layer just outside the pipe, the layer of fine nichrome heating wire wound around the insulating layer, the other electric-insulating thin layer over the nichrome wire, and the thick thermal insulator of about 80-mm thickness as the outside layer. Thermocouples for measurement of the pipe-wall temperature are attached at 10 points of the outer surface of the heated pipe in intervals of 500 mm. The pipe is heated under the condition of constant heat flux by the nichrome wire and in the fully developed temperature-field region the temperature gradient in the axial direction is constant, as shown in Fig. 11.2. The air flow rate of about 3 kg/h is measured by the orifice flow meter. The Reynolds number ($2W_m a/v$) region reaches up to about 3000 as the critical Reynolds number increases gradually with the Rayleigh number ($g \beta \Delta T\, a^3/\kappa v$). The maximum average velocity is of the order of 1.5 m/s. The symbols used in the Reynolds and Rayleigh numbers are defined subsequently in Notation.

A Pitot total pressure probe and a static pressure probe having a small orifice are put together with the thermocouple as shown in Fig. 11.3. The

Fig. 11.2. Temperature profile of heated pipe wall.

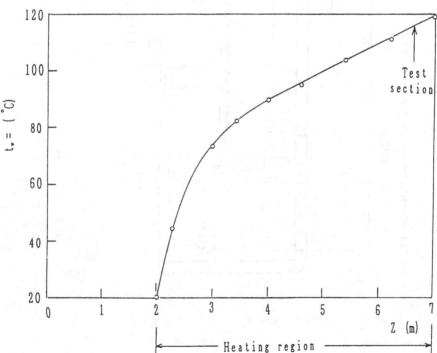

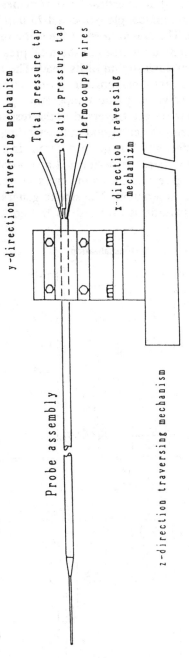

Fig. 11.3. Probe assembly and traversing device.

probes are assembled in a body and fitted in a supporting arm to be traversed in the horizontal (*x*-axis) and vertical (*y*-axis) planes as seen in Fig. 11.1(b). The junction of the thermocouple is located *15 mm* behind the head of the total pressure probe. The closed head of the static pressure probe is fixed in the assembly 5 mm behind the head of the total pressure probe and its static pressure hole is 5 mm downstream of its head. The total and static pressure difference is measured by a manometer having an accuray of 0.02 Pa.

In measurements of velocity and temperature profiles in horizontal and vertical planes, the probes are traversed in the respective planes. In the last stage of the experiment, to get a good understanding of the flow field of the combined heat transfer, which is quite different from the Poiseuille flow, a flow visualization experiment is performed at a location 320 mm upstream from the pipe exit. Figure 11.4 shows the cross-sectional view of the part for the flow visualization experiment. The smoke generating device is shown in Fig. 11.1(a). Liquid paraffin is dropped on the hot nichrome wire and the

Fig. 11.4. Cross-sectional view of flow visualization.

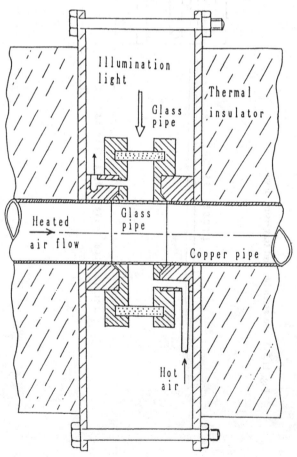

smoke of paraffin thus generated is introduced into the air flow. The illumination light is introduced vertically to the pipe through the glass of the flow visualization section. As the inside of the pipe other than the visualization section is dark enough, the secondary flow generated by buoyancy is thus visualized as a smoke pattern and can be photographed with an exposure time of about 1/25 s. Figure 11.5 is one of the photos taken using the process just explained.

The test section is 340 mm upstream of the pipe exit and the flow visualization section is located 30 mm downstream of the test section.

Procedure

When the motor of the air blower is switched on, the air flow rate is set at about 3 kg/h. The nichrome heater is also turned on. (These two

Fig. 11.5. Photograph of secondary flow due to buoyancy forces in a uniformly cooled horizontal straight tube. $Re = 2 \times 10^3$, $Ra = 10$, $Pr = 0.71$ $Re \cdot Ra = 2 \times 10^4$, $L/(dRePr) = 0.077$ (L = cooling length, d = inner diameter, dotted lines = stream lines of the secondary flow computed by numerical analysis).

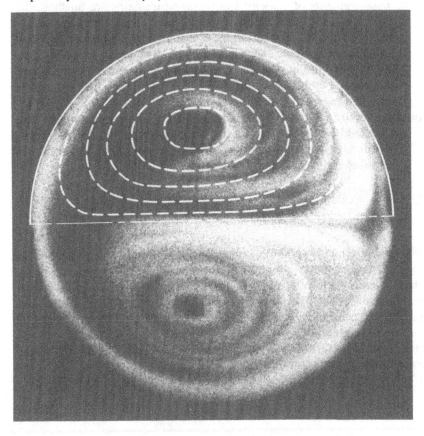

procedures should be started about one hour in advance of the student laboratory class as it takes time to warm the insulator layer of
the pipe to reach the steady-state temperature.)

The pipe has to be heated to attain a constant axial temperature gradient of
about 10 °C/m in the developed flow region and in the test section.

The 6-mm diameter probe assembly is installed in the fitting hole of the
probe-traversing device shown in Fig. 11.3. The orientation of the
axis of the probe assembly is adjusted to oppose the air flow ($-z$
direction) and lie in the horizontal axis. Next, the probe assembly is
introduced into the pipe along the pipe axis by the traversing device
in the negative z direction until the head of the Pitot tube reaches
the test section.

The total pressure and static pressure probes are connected to the pressure
difference manometer, and the thermocouple probes are connected
to the millivoltmeter.

In the test section, the probes are traversed from the center in the radial
direction, and the horizontal (x direction) and vertical (y direction)
planes. Measurements of velocity and temperature are made at
3-mm intervals.

After the measurements of velocity and temperature fields are finished, the
probe assembly is taken out of the pipe and removed from the traversing device. Then, the lamp is turned on to illuminate the flow
visualization section. The smoke-generating heater is started and
the valve of the liquid paraffin container is opened. The secondary
flow caused by buoyancy is shown in Fig. 11.5. The flow is observed
from the pipe exit side.

It should be noted that the temperature profile is quite different from that
expressed by the fourth-order algebraic equation found in any heat-
transfer textbook. Large heat-transfer enhancement is seen due to
combined convective heat transfer.

Explanation

The experiments and observations reveal that the convective field in a heated
horizontal circular pipe is influenced enormously by the strong secondary
flow that is generated by buoyancy of a hot fluid heated at the inner surface of
a pipe. The convective field rises in the boundary layer along the surface and
descends in the central core of the flow. In the boundary layer, the integral
momentum equation in the circumferential direction which includes the
buoyancy term together with the integral equation of energy reduce the mean
thickness of the momentum boundary layer δ_m to

$$\delta_m = C \left[\frac{\zeta}{ReRa} \right]^{1/5} \tag{11.1}$$

where C is a constant. ζ is the ratio of the momentum to the thermal boundary layer thicknesses and is a function of Prandtl number.

The temperature gradient in the axial direction near the measuring section is constant. It is denoted by τ. The heat flux q is given as

$$q = a\gamma C_p \, W_m \, \tau/2 \qquad (11.2)$$

where γ and C_p are the specific weight and isobaric specific heat of the fluid, respectively. The product of the specific weight and axial mean velocity W_m is obtained from the measured weight flow rate.

With respect to the coordinates in Fig. 11.1(b), the small area of a cell in a cross section is denoted by $\Delta x_i \, \Delta y_i$. When the measured axial velocity and temperature in the cell are expressed by W_{ij} and t_{ij}, respectively, the bulk temperature is given by

$$t_b = \sum_{ij} W_{ij} t_{ij} \Delta x_i \Delta y_i \Big/ \sum_{ij} W_{ij} \, \Delta x_i \Delta y_j \qquad (11.3)$$

Figures 11.6 and 11.7 show measured velocity and temperature distributions in the vertical and horizontal planes, where t_c and t_w are the temperatures at the center and the wall in a cross section, and W_c is the air velocity at the center. To calculate W_{ij} and t_{ij} in a cross section with an angle θ to the horizontal plane, the values of W_{ij} and t_{ij} from the highest part to the lowest part in the vertical plane in the main flow are assumed to follow a cosine function. The heat transfer coefficient α is defined as

$$\alpha = q/(t_w - t_b) \qquad (11.4)$$

Nusselt number Nu is given by $Nu = 2\alpha r/\lambda$, where λ is the thermal conductivity of the fluid.

Figure 11.8 shows the results of experiments described herein plotted in terms of $Re \cdot Ra \cdot Pr$. The vertical axis is the ratio of Nu obtained by theory

Fig. 11.6. Temperature and velocity profiles in the horizontal direction.

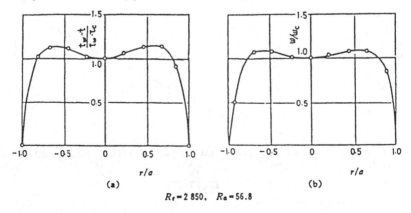

(a)   (b)

$R_r = 2\,850, \quad R_a = 56.8$

and experiment over the Nusselt number without the secondary flow; that is, $Nu_o = 48/11$. The solid line expresses that obtained by boundary-layer theory, and the broken line is numerically calculated. The large value of the ratio at high $Re \cdot Ra \cdot Pr$ indicates a typical example of large heat-transfer enhancement of the combined convective heat transfer due to the secondary flow, where Pr is the Prandtl number defined as ν/κ.

Fig. 11.7. Temperature and velocity profiles in the vertical direction.

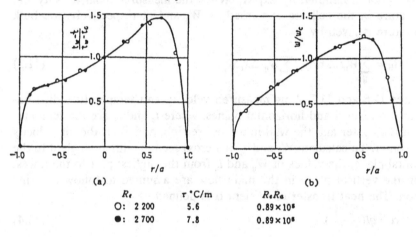

R_e	τ °C/m	$R_e R_a$
O: 2 200	5.6	0.89×10^8
●: 2 700	7.8	0.89×10^8

Fig. 11.8. Nusselt-number performance.

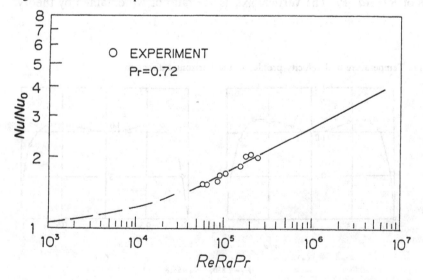

Suggested headings

Data: $a = 30$ mm; $P_{static} =$ nearly atmospheric; $t =$ less than 120 °C
       $z =$ distance from the heating start point
       $v =$ _____; $\kappa =$ _____; $\gamma =$ _____ ; $C_p =$ _____

Coordinates		Measured quantities						Calculated quantities			
x_{ij}	y_{ij}	W_{ij}	t_{ij}								
z				W_c	W_m	t_c	t_w	t_b	q	Nu	Ra

References

1. Mori, Y.; Futagami, K.; Tokuda, S., and Nakamura, M., "Forced convective heat transfer in uniformly heated horizontal tubes (1st report)," *Int. Journal of Heat and Mass Transfer* 9 (1966): 453–563.
2. Mori, Y., and Futagami, K., "Forced convective heat transfer in uniformly heated horizontal tubes (2nd report)," *Int. Journal of Heat and Mass Transfer* 10 (1967): 1801–13 .

Notation

W_m	mean axial velocity
a	radius of the pipe
v	kinematic viscosity
g	gravitational acceleration
β	expansion coefficient of air
ΔT	difference between wall and mean air temperatures in cross section
κ	thermal diffusivity of air

Yasuo Mori

Yasuo Mori was educated at the University of Tokyo (D. Eng.), after which he spent 5 years at the Physico-Chemical Research Institute and then 30 years at the Tokyo Institute of Technology. His research and education cover thermodynamics and heat transfer, specifically convective heat-transfer enhancement by secondary flow and radiation, plasma, and electro-hydrodynamic heat transfer. He is chairman of several national energy projects in Japan.

EXPERIMENT 12
Measurement of convective heat-transfer coefficients on external surfaces

Contributed by
ROBERT F. BOEHM

Principle

The lumped-mass assumption (negligible internal thermal resistance) is used to infer forced-, natural-, or mixed-convection heat-transfer coefficients on spheres, cylinders, and other shapes.

Objective

This experiment allows the estimation of heat-transfer coefficients that result from external flows. Use of the lumped-mass approximation is a key element to the work, and this experiment can be used to explore the limits of this important experimental convective technique. Also of value is the estimation of the radiation contribution compared to the convective contribution in the total heat loss from a heated object. The basic approach can be used for a variety of geometries in forced-, free-, or mixed-convection arrangements. The description here focuses on *forced-convection* applications, but the basic apparatus and concept can be used for the other situations as well.

Apparatus

Constant-speed centrifugal fan with uniform flow outlet and damper on inlet (for forced-flow experiments)
Hand-held anemometer, propeller type (for forced-flow experiments)
Thermocouple reference junction and signal readout device (many data loggers combine these functions and can be used). A highly desirable alternative is a computer-based data acquisition system.
Clock (if time is not recorded with data logger or computer)
Support stand
Bunsen burner and lighter
Two thermocouples, one in ambient air and one with adaptor for test element

Barometer

Copper test elements (e.g., sphere, cylinder)

A diagram of the test apparatus suitable for forced-flow experiments is shown in Fig. 12.1. Here, the flow environment is furnished by a low-cost centrifugal fan that has been modified to allow flow control and a uniform output velocity field. We use a simple bolt-on damper to allow flow control. A bundle of common drinking straws on the exhaust of the fan is used for assuring a uniform velocity field. However, any method might be used that will assure a uniform flow in a cross-sectional area at least three times as large as the test element. If possible, the velocity field should be mapped once with a point velocity measuring device to assure that the field is reasonably uniform. The damper should function such that at least an order of magnitude in velocity turndown is possible. A larger turndown range, such as might be available with a variable speed blower, or by using two different blowers, is a highly desirable alternative. An anemometer should be used that will indicate an average velocity over an area slightly larger than the frontal area of the test element.

The test element, both a sphere and a cylinder are shown in Fig. 12.2,

Fig. 12.1. The experimental apparatus. Not shown are temperature- and pressure-measurement devices for the ambient air.

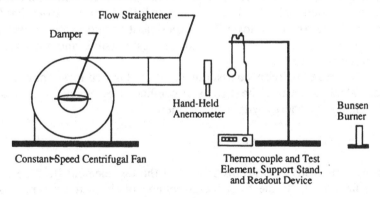

Constant-Speed Centrifugal Fan

Thermocouple and Test
Element, Support Stand,
and Readout Device

Fig. 12.2. Cross sections of possible test elements.

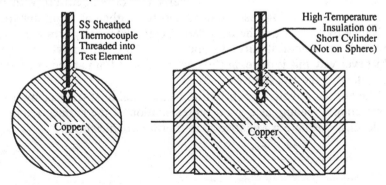

should be constructed of as large a size as possible without having the profile exceed one-third of the flow stream area and without negating the lumped-mass requirements. Elements approximating two-dimensional flow situations (such as the "infinite" cylinder and denoted here as 2D elements) must have their ends insulated. In addition, positioners fastened to the insulation may be required to assure that the 2D elements remain perpendicular to the flow stream. A bare-tipped, stainless-steel sheathed thermocouple is threaded and inserted into the test element. This thermocouple is used both for temperature readout as well as for support of the element. The tip of the thermocouple must contact the copper at the bottom of the blind-tapped hole.

Procedure

Heat the test element to about 600 °C with the flame from the Bunsen burner. Simultaneously turn on the fan, and set a desired fan velocity by adjusting the damper. Check and record the velocity with the anemometer. Quickly move the heated test element from the Bunsen burner flame into the air flow stream. Record the temperature of the element at regular time intervals. Actually only a few readings at the beginning of the test and a few when the test element is about 100 °C might be used. After the test element has cooled, remove it from the air stream, measure the air velocity, and compare this value to that found initially. There should be little difference from that found before and after the cool down. The experiment should be performed for at least three air velocities: maximum value, lowest possible, and one which falls in between the extremes.

The data are to be reduced assuming that the lumped-mass approximation holds. Calculate the total heat transfer coefficient by equating the heat loss to the change of energy within the element.

$$hA_e(T - T_\infty) = -\rho_e c_e V_e \frac{dT}{dt} \tag{12.1}$$

Here A_e is the active heat-transfer area of the test element (not the area of any insulation), V_e is the total metal volume of the test element (not the volume of any insulation), c_e is the specific heat of the element, and ρ_e is the density of the element. Estimate the derivative of temperature with respect to time for the cooling test element. Do this either by approximating the derivative by $\Delta T/\Delta t$ from the recorded discrete data, or by fitting the discrete data with an analytical curve fit and then taking the analytical derivative. Whichever way this is done, at the minimum do it both at the beginning and the ending of the transient cool down.

Estimate the radiative contribution to the total heat transfer by using the thermocouple reading as a good representation of the surface temperature. To do this, calculate a radiative heat-transfer coefficient h_r.

$$h_r = \varepsilon_e \sigma \frac{(T^4 - T_\infty^4)}{(T - T_\infty)} \tag{12.2}$$

This value will be compared to the value for the combined heat transfer coefficient found from Eq. 12.1. For low values of emissivity, say for polished copper, and for larger values of velocity, the radiation will normally not be important. This may not be the case for natural or mixed convection.

Now check the validity of the lumped-mass assumption. Calculate the Biot number for the test element. Use the total heat-transfer coefficient with the thermal conductivity (k_e) of the test element as follows:

$$Bi = \frac{h\,V_e/A_e}{k_e} \tag{12.3}$$

If Bi is less than about 0.1, the lumped-mass assumption holds.

Cast the data for the convective heat-transfer coefficient ($h_c = h - h_r$) into nondimensional form and compare to any data correlations found in the literature. For forced air flow over a sphere, one correlation is given by McAdams:[1]

$$Nu_f = 0.37 Re_f^{0.6}, \quad 17 < Re_f < 70{,}000 \tag{12.4}$$

Here,

$$Nu_f \equiv \frac{h_c D}{k_f} \quad \text{and} \quad Re_f \equiv \frac{U D}{v_f}$$

and the subscript f denotes air properties evaluated at the film temperature. Another correlation has been reported for spheres more recently by Whitaker:[2]

$$Nu_\infty \equiv \frac{h_c D}{k_\infty} = 2 + (0.4\,Re_\infty^{1/2} + 0.06\,Re_\infty^{2/3})\,Pr^{0.4}\left(\frac{\mu_\infty}{\mu_s}\right)^{1/4} \tag{12.5}$$

Air properties here are evaluated at the ambient conditions, except for μ_s, which is the viscosity evaluated at the test element temperature. Restrictions on this latter correlation are $3.5 < Re_\infty < 76{,}000$, $0.71 < Pr < 380$, and $1.0 < \mu_\infty/\mu_s < 3.2$. Note that this latter restriction cannot be satisfied in this experiment because the surface temperature of the test element is always higher than that of the bulk air. However, comparison will be made to this equation in spite of this limitation. These correlations are compared in Fig. 12.3.

Data taken from experiments correlate quite closely with the relations shown in Fig. 12.3. Locations of the data will obviously depend upon the velocity range possible from the fan used.

Additional Comments

1. We have performed the experiment described here with 25.4-mm (1-inch) diameter copper spheres with good results. Results using 12.4-mm diameter by 25.4-mm long cylinders have yielded a less satisfactory outcome. See more on this later.
2. Little difference should be found between the results of the data taken at the beginning of the transient and that found at the end of the transient

cool down for the same air velocity. The only items affected are the properties (really the viscosity) and the thermal radiation.

3. An alternative for any of the experiments described hitherto is to use a steady-state technique, which requires an imbedded heater in the test element. Here the lumped-mass approximation would be important to ensure a uniform temperature throughout the test element if temperature is measured at only one point.

4. The use of short cylinders is problematic when used as described here. To approximate the 2D situation, the ends must be effectively insulated with a material that can tolerate flame. Most of these kinds of insulations are quite dense, possibly adding a significant amount to the thermal capacity of the test element. Probably the steady-state approach noted in item 3 will be best applied if short cylinders and related geometries are used with insulated ends. It is possible to evaluate short cylinders or bars with bare ends and attempt to combine results from simpler geometry correlations (say flat plates and very long cylinders), but this approach may not yield good agreement with standard correlations.

5. We have also had very satisfactory outcomes using vertical aluminum plates in natural-convection laboratory exercises. Since the heat transfer coefficients in natural convection in air are generally much less than those in the forced-flow situations, the lower thermal conductivity of aluminium (compared to copper) still allows the lumped-mass assumption to be applied. As the plate cools, both the Grashof number and the Nusselt number decrease, allowing a range of data to be gathered in one run. The range of both values can also be affected by changing the vertical dimension of the aluminum plates. This allows the coverage of a wide range of the natural-convection correlations. Depending upon the amount of oxidation on the aluminum, radiation can be a dominant factor in the natural-convection experiments of the type discussed here. Also, any spurious air currents (say from room heating/cooling systems) can have profound and undesirable effects on the results.

Fig. 12.3. Comparison of the correlation recommended by McAdams[1] and the correlation given by Whitaker.[2] The viscosity term in the latter equation is taken to be unity here.

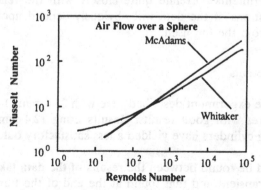

6. One area of concern with the approach described here is conduction errors in the support (the thermocouple in Figs. 12.1 and 12.2). It is probably wise to have students estimate the amount of this error.

Suggested headings

Constants:

Element _____ ; dimension(s) = _____ ; ε_e (estimate) = _____

V_e = _____ ; A_e = _____

Each subset of data is characterized overall by flow velocity.

U = _____ ; P_∞ = _____ ; T_∞ = _____

Recorded/computed data:
t T c_e k_e ρ_e T_f v_f h_r h h_c Bi Re_f Nu_f

Comparisons are then made to the correlations given.

References

1. McAdams, W., *Heat Transmission*, third ed., McGraw-Hill, New York, 1954.
2. Whitaker, S., "Forced convection heat transfer correlations for flow in pipes, past flat plates, single cylinders, single spheres, and for flow in packed beds and tube bundles," *AIChE Journal* 18, 1972: 361–71.

Robert F. Boehm

Robert Boehm received his Ph.D. in mechanical engineering from the University of California at Berkeley in 1968. He was at the University of Utah, Department of Mechanical Engineering, until 1990, serving as department chairman from 1981 to 1984. He is now Chairman of the Department of Mechanical Engineering at the University of Nevada, Las Vegas. His work has been primarily in thermal-systems design and heat-transfer problems applied to energy and bioengineering.

EXPERIMENT 13

Measurement of local heat-transfer coefficient on the ice surface around isothermally cooled cylinders arranged in a line

Contributed by
TETSUO HIRATA

Principle

At the ice–water interface in a steady-state condition, the heat flux transferred from water to the ice–water interface is equal to that conducted from
the ice–water interface to ice. By measuring the coordinates of the ice–water
interface, the heat flux from the interface to ice is calculated by the boundary-
element method. The local heat-transfer coefficient on the ice–water interface is, then, estimated by Newton's law of cooling.

Object

Ice formation around tubes in a water flow relates to many practical problems such as lowering thermal efficiency or increasing pressure drop in a
water-cooled heat exchanger in a refrigeration system. It also relates to many
other applications such as the ice-bank method in a low-temperature heat-
storage system. In those cases, the local heat-transfer coefficient on the ice
surface is an important factor in predicting the ice amount around the tubes
and also the thermal efficiency of the heat exchanger. The measurement
of the local heat-transfer coefficient, therefore, presents essential information
for practical designs.

Apparatus

The experimental apparatus consists of a calming section, a test section, a
flow meter, a refrigeration unit, and two circulation systems of water and a
coolant as shown in Fig. 13.1. In Fig. 13.2, a schematic illustration of the test
section is shown. The test section has a 0.15-m × 0.04-m cross-sectional area
and has a 1.0-m length. The walls are made of transparent acrylic resin plates
in order to observe the growth of the ice layer, and are installed in the
vertical position to minimize the effect of the natural convection of water.
Ten isothermally cooled cylinders with 0.041-m o.d. and 0.04-m length shown

in Fig. 13.3 are arranged in a line and installed in the test section. To produce a uniform wall temperature, copper is used to fabricate the cylinders. The coolant is led directly from a constant-temperature bath to each cylinder and is circulated at a high velocity inside the cylinders. The flow of the water in the test section is in the upward vertical direction. The two dimensionality of heat flow from the water to the cylinders is confirmed by the fact that a two-dimensional ice layer is produced around the cylinders. Although the effect of heat conduction through the acrylic resin plates of the test section wall is observed at both ends of each cylinder, the region with this effect is considered to be negligible compared to the cylinder length.

Fig. 13.1. Experimental apparatus.

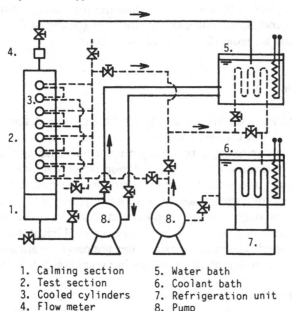

1. Calming section
2. Test section
3. Cooled cylinders
4. Flow meter
5. Water bath
6. Coolant bath
7. Refrigeration unit
8. Pump

Fig. 13.2. Schematic representation of ice layer in the test section (Ref. 4, p. 707).

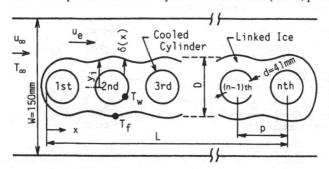

Procedure

The wall temperatures of the cooled cylinders T_w are evaluated from two thermocouples located at the top of and 90 deg to the third and fifth cylinders, respectively. Although a small temperature difference between the two cylinders is detected during the initial stage of ice formation, the difference becomes negligible with ice growth. The free-stream water temperature T_∞ is estimated from the mean value of the inlet and outlet of the test section. An isothermal wall condition is satisfied within an error of 3 percent in T_w. The inlet water temperature variations due to the dynamics of the refrigeration system yield a maximum error of 2 percent in T_∞. After a steady-state condition is reached, a small amount of dye is injected into the water stream to enhance the ice–water interface. To measure the thickness of the ice layer, the ice around the cylinders is photographed and the coordinates of the ice–water interface are measured. An anastigmatic lens (Nikon Micro Nikkor 200-mm F4S) is used to avoid distortion of the subject.

Explanation

For the steady-state ice layer around a single cylinder, some determination methods for the local heat-transfer coefficient have been reported, such as a heat conduction analysis using a boundary-fitted coordinate system[3] and a point-matching technique using an exact solution of the Laplace equation for polar coordinates.[2] In this study, the cylinders become linked by an ice layer under conditions of a thicker ice layer, so a completely different shape of ice from that for a single cylinder is produced. Therefore, the previous determination methods for the local heat transfer coefficient cannot be applied. The other method, which measures the temperature gradient in a water stream on the ice–water interface, results in a comparatively large error in measurement.

In the present measurement, the boundary-element method[1] is used to determine the local heat transfer coefficient on the surface of the ice. For this method, the data necessary are the coordinates of the ice–water interface and

Fig. 13.3. Schematic illustration of cooled cylinder.

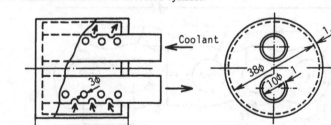

the temperature boundary conditions. Figure 13.4 shows three patterns of ice shape expected to form around the cylinders. The effect of the number of boundary elements, m, on the accuracy of the solution is checked for the ice pattern shown in Fig. 13.4(c). In Fig. 13.5, the effect of m on the local heat transfer results are shown. It is ascertained that the effect of m is less than 5 percent for $m > 56$. The temperature boundary condition at the ice–water interface and cylinder wall is a uniform and constant temperature. At the other boundaries, a zero temperature gradient is used. The temperature gradient inside the ice layer is computed by the boundary-element method. The local heat transfer coefficient h_x on the ice surface in a steady-state condition is determined by

$$h_x(T_\infty - T_f) = \lambda_i \frac{\partial T}{\partial n}\bigg|_{y=y_i} \tag{13.1}$$

where T_f is the freezing temperature of water, λ_i is the thermal conductivity of ice, and y_i is the coordinate of the ice–water interface. The measurement of the coordinates of the ice–water interface yields a maximum error of 3 percent in the amount of ice and it produces an error of about 8 percent in h_x. Figure 13.6 shows the typical results of local Nusselt number versus Reynolds number defined by Eq. (13.2).

$$Nu_x = \frac{h_x x}{\lambda_w}, \quad Re_x = \frac{u_e x}{v_w} \tag{13.2}$$

It is shown that Nu_x oscillates with increasing Reynolds number with the average value of Nu_x increasing steadily, and that the tendency of the average value of Nu_x versus Re_x is the same as that for turbulent heat transfer on a flat plate.

Fig. 13.4. Typical ice patterns formed around the cylinders (Ref. 4, p. 709).

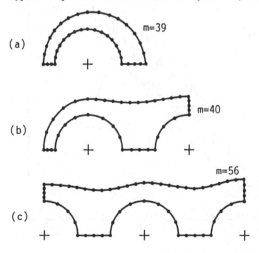

Fig. 13.5. Effect of element number for the boundary-element method.

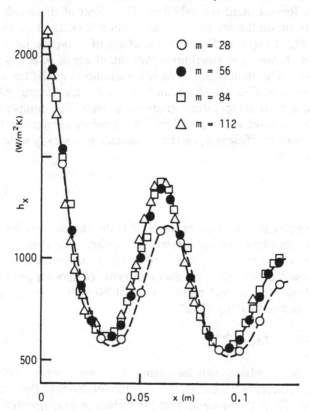

Fig. 13.6. Typical result of local Nusselt number (Ref. 4, p. 710).

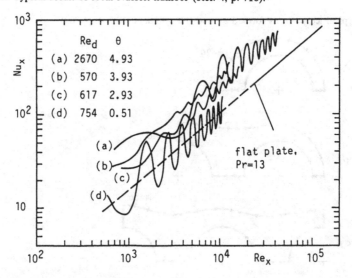

Suggested headings

Parameter ranges: $p = 60–100$ mm; $Re_d = 500–3000$; $T_\infty = 1.0–4.0$ °C; $u_\infty = 0.02–0.11$ m/s; $\theta = 0.5–7.0$

References

1. Brebbia, C. A., *The Boundary Element Method for Engineers*, Pentech Press, London, 1978.
2. Cheng, K. C.; Inaba, H., and Gilpin, R. R., "An experimental investigation of ice formation around isothermally cooled cylinder in crossflow," *ASME J. Heat Transfer* 103 (1981): 733–8.
3. Cheng, K. C., and Sabhapath, P., "Determination of local heat transfer coefficient at the solid–liquid interface by heat conduction analysis of the solidification region," *ASME J. Heat Transfer* 107 (1985): 703–6.
4. Hirata, T., and Matsui, H., "Ice formation and heat transfer with water flow around isothermally cooled cylinders arranged in a line," *ASME J. Heat Transfer* 112 (1990): 707–13.

Notation

d	diameter of cylinder
h	local heat transfer coefficient
Nu_x	local Nusselt number
Re_d	cylinder Reynolds number, ud/ν_w
Re_x	Reynolds number
T_f, T_w, T_∞	temperature of freezing, cylinder wall, and water
u_e, u	velocities of water around cylinders with and without ice
u_∞	free-stream velocity of water upstream of cylinders
y_i	distance from center axis of row of cylinders to ice–water interface
θ	dimensionless cooling temperature ratio, $(T_f - T_w)/(T_\infty - T_f)$
λ_i, λ_w	thermal conductivities of ice and water
μ_w	kinematic viscosity of water

Tetsuo Hirata

Professor Hirata is an associate professor of mechanical systems engineering at Shinshu University, Nagano, Japan. He received a D.Eng. degree (1977) in mechanical engineering from Hokkaido University, Sapporo. From 1977 to 1979, he was a post doctoral fellow at the University of Alberta, Canada. His current research is concerned with conduction–convection mixed heat transfer in phase-change problems.

EXPERIMENT 14

Experiments of unsteady forced convection in ducts with timewise variation of inlet temperature

Contributed by
W. LI *and* S. KAKAÇ

Principle

The temperature response of the unsteady inlet temperature varying with time is an essential parameter in the thermal design of the heat exchanger and other heat transport equipment. The temperature amplitude variation along the general passage decays exponentially and can usually be determined by experiment. This variation will affect the entire heat transport process within the heat exchanger and other heat-transport equipment.

Objective

The objective of this experiment is to determine the decay of the temperature oscillation along the channel for a timewise oscillation of inlet temperature. In practical applications, the heat transfer within the channel may be exposed to some planned or unplanned transients or start-ups and shutdowns during the operation. Thus, such a knowledge is critically necessary for those devices which never attain steady-state operation because of their nature of periodical operation in time.

Apparatus

The experimental apparatus consists of a rectangular duct with different sections of filter, calming, inlet, test, and convergence. The geometry of the test section is a rectangular duct with a cross section of 254×25.4 mm^2 (10×1 in.^{2}). The instrumentation includes a wave generator, a power supply, a heater, an inclined manometer, voltmeters, thermocouples, an orifice plate, and a fan, as shown in Fig. 14.1.

Air flows from the calming section to the inlet section (2770 mm in length) wherein the velocity becomes fully developed. The duct is constructed with an outer casing made from 6.35-mm (1/4-in.) thick plywood with outer dimensions of 114.3 mm $\times$ 381 mm $\times$ 4670 mm which incorporates the inlet and

Fig. 14.1. Schematic diagram of experimental set up.

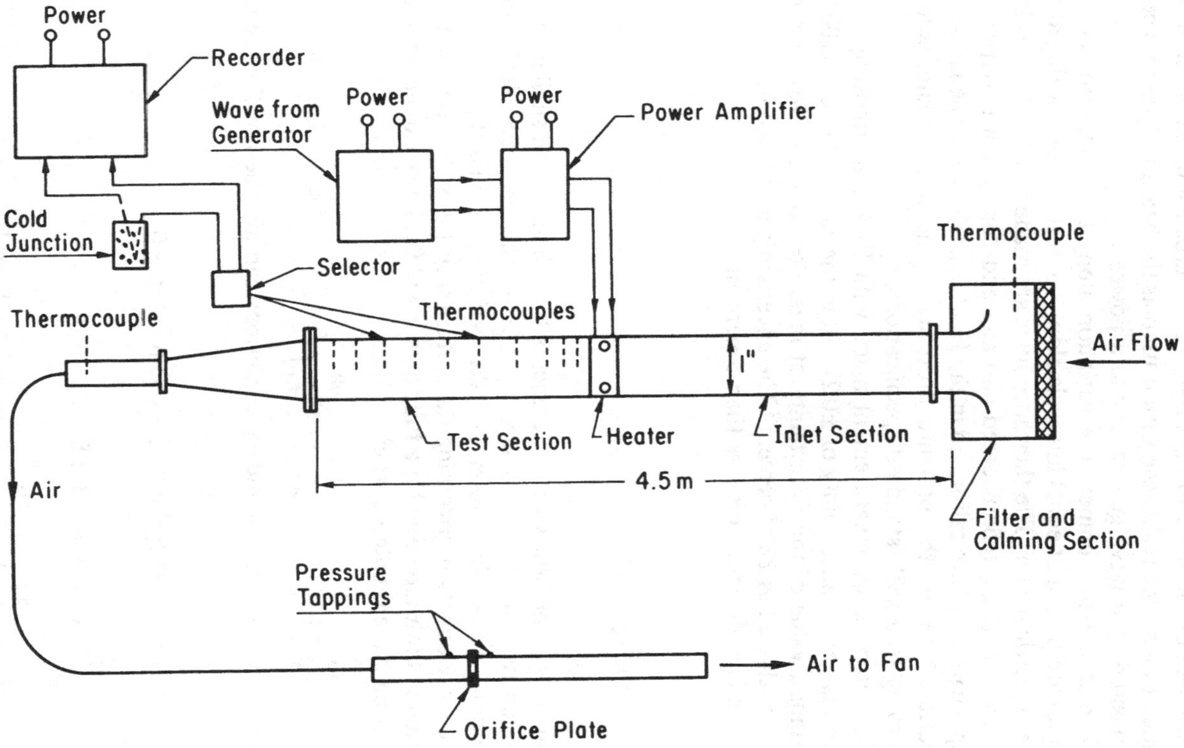

test sections. The interior of the casing is lined with 25.4-mm thick extruded styrofoam leaving a cross-sectional flow area of 254×25.4 mm^2 (10×1 in.^{2}).

Throughout the study, periodically varying heat input is provided by an electric heater. In order to minimize the disturbance to the air flow, 0.4-mm diameter nichrome resistance wire is used as the heating element. The heater is powered by a wave generator and a power supply.

To measure the temperature variation along the duct, fourteen thermocouples are placed at equal intervals along the test section starting at the exit of the electric heater. The thermocouples are made from 0.01-in. (0.254-mm) diameter, 30 gage, teflon-coated chromel and constantan (E-type) thermocouple wires, and are calibrated in the Heat Transfer Lab. The mass flow rate is calculated from the pressure drop across the orifice plate, constructed according to ASME standard specifications.[1]

The temperature measurement made with E-type thermocouples usually has a maximum uncertainty of $\pm 0.2°$ within the range of 0 °C to 100 °C. The maximum value of the uncertainty of mass flow rate (or Reynolds number Re) is evaluated as ± 2.64 percent.[2] The uncertainty associated with decay index (or standard deviation) is less than 5 percent.

Background

Assume that the flow within the duct is two-dimensional, similar to the flow between two parallel plates separated by a distance of $2b$. Neglecting axial diffusion, viscous dissipation, and the variation of fluid thermal properties, the energy equation governing the diffusion in the y direction and the convection in the x direction for a fully developed fluid flow can be written in the following dimensionless form:[3]

$$\frac{\partial \theta}{\partial \tau} + U(\eta)\frac{\partial \theta}{\partial \xi} = \frac{\partial}{\partial \eta}\left(\varepsilon(\eta)\frac{\partial \theta}{\partial \eta}\right), \text{ for } \xi > 0, 0 < \eta < 1, \tau > 0 \qquad (14.1)$$

and the inlet and the boundary conditions can also be nondimensionalyzed as:

$$\theta(0,\eta,\tau) = \Delta\theta(\eta)\,e^{i\Omega\tau}, \text{ for } 0 < \eta < 1, \tau > 0 \qquad (14.2)$$

$$\frac{\partial \theta}{\partial \eta} = 0, \text{ at } \eta = 0, \text{ for } \xi > 0, \tau > 0 \qquad (14.3)$$

$$Bi\,\theta + \frac{\partial \theta}{\partial \eta} + \frac{1}{a^*}\frac{\partial \theta}{\partial \tau} = Bi\,\theta_\infty, \text{ at } \eta = 1, \text{ for } \xi > 0, \tau > 0 \qquad (14.4)$$

By the method of separation of variables, we can separate the dimensionless temperature distribution $\theta(\xi,\eta,\tau)$ into two parts, one corresponding to the steady distribution $\theta_1(\xi,\eta)$, the other one representing the periodic oscillation $\theta_2(\xi,\eta,\tau)$.

$$\theta(\xi,\eta,\tau) = \theta_1(\xi,\eta) + \theta_2(\xi,\eta,\tau) \tag{14.5}$$

The solutions of $\theta_1(\xi,\eta)$ and $\theta_2(\xi,\eta,\tau)$ can be obtained by extending the generalized integral transform technique.[2] The solution of $\theta_1(\xi,\eta)$ is

$$\theta_1(\xi,\eta) = \theta_\infty\left(1 - \sum_{n=1}^{\infty} \frac{1}{\sqrt{N_n}} C_n e^{-\lambda_n^2 \xi} Y_n(\eta)\right) \tag{14.6}$$

and the solution of $\theta_2(\xi,\eta,\tau)$ can be expressed as

$$\theta_2(\xi,\eta,\tau) = e^{i\Omega\tau} \sum_{k=1}^{N} g_k(\eta) e^{-\mu k \xi} \tag{14.7}$$

The amplitude of the temperature oscillation $\theta_{amp}(\xi,\eta)$ is defined as

$$\theta_{amp}(\xi,\eta) = \sqrt{\left[Re(\theta_2 e^{-i\Omega\tau})\right]^2 + \left[Im(\theta_2 e^{-i\Omega\tau})\right]^2} \tag{14.8}$$

Procedure

For each data run, both temperature oscillation frequency (β) at inlet and the Reynolds number (Re) are fixed, but they are adjustable in the entire experiment. First, the pressure drop across the orifice plate can be adjusted to the desired range, and the inlet frequency can be stabilized on the selected value. After the electric heater is turned on, temperature amplitudes at various locations should be checked until they are not changing with time. Then, at a fixed value of Reynolds number and a given inlet frequency, the oscillation of temperature along the duct as a function of time and the temperatures before and after the orifice plate can be recorded by thermocouples; see Fig. 14.2.

After the completion of data recordings (temperatures and pressure drop), the inlet frequency is reset to the next desired value. All temperatures and the pressure drop will be remeasured until the steady temperature amplitude is reached. After experiments for all desired inlet frequencies are carried out, the experiments for another desired pressure drop across the orifice plate (Reynolds number) will be repeated.

In the experiments, the temperatures can be converted from the measured thermocouple's voltage through the calibration chart. The Reynolds number of the flow can be calculated from the mass flow rate and properties of air and other geometries as

$$Re_{D_e} = \frac{m}{\pi\mu D_e/4} \tag{14.9}$$

while the mass flow rate can be found from the following relation:

$$m = C\sqrt{\rho_{air}\Delta P} \tag{14.10}$$

where C is a constant for a specific orifice plate, which can be determined from the geometries of the orifice plate, ρ_{air} is the density of the air, and ΔP is the pressure drop across the orifice plate.

During the experiments, when the temperature at the entrance is specified as a sinusoidal oscillation, as shown in the recordings (Fig. 14.2), the temperatures along the duct are changing sinusoidally with the same frequency as at the inlet. Furthermore, the amplitude of oscillations is decaying quite rapidly at positions away from the entrance. Since the primary concern of the present study is to measure the decay of the temperature variation, the temperature amplitudes are further discussed.

The amplitude of temperature oscillations on the centerline of the duct is obtained from the maximum and minimum values of the temperature readings (T_{max} and T_{min}) as

$$\Delta T = (T_{max} - T_{min})/2 \tag{14.11}$$

Fig. 14.2. The samples of temperature variations at different locations along the ducts.
(a) $Re = 491$, $\beta = 0.01$ Hz, paper speed = 0.50 mm/s in laminar flow.
(b) $Re = 15{,}902$, $\beta = 0.01$ Hz, paper speed = 0.25 mm/s in turbulent flow.

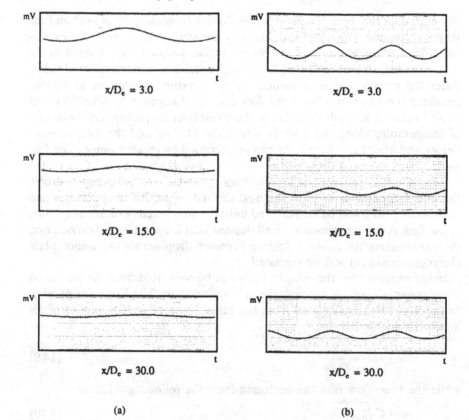

(a)   (b)

In Figs. 14.3 and 14.4, the temperature amplitudes at different locations are plotted against dimensionless distance x/D_e for different inlet frequencies β at selected Reynolds numbers Re in the laminar-flow regime. Figures 14.5 and 14.6 are the examples of the experimental results for turbulent flow. Except the points close to the inlet, the experimental data almost fall on a straight line in these semilog plots. Regression lines of the experimental results are drawn in the figures. The temperature amplitude variation may be expressed as:

$$\Delta T_c/\Delta T_i = e^{-\alpha x/D_e} \qquad (14.12)$$

The decay index α is directly given by the slope of the curve. The values of the temperature amplitude decay index for different Reynolds number both in laminar and turbulent regimes are listed in Table 14.1 and Table 14.2.

Results

The effect of inlet frequency on the variation of the temperature amplitude can be observed from semilog plots. The linear appearance of the experimental results in semilog coordinates is due to the exponential decay of the amplitude along the duct, which has been illustrated in Eq. (14.7).

Generally, the temperature amplitude decays exponentially along the duct

Fig. 14.3. Decay of temperature amplitude along the duct in laminar flow for $Re \approx 660$ and $Pr \approx 0.70$.

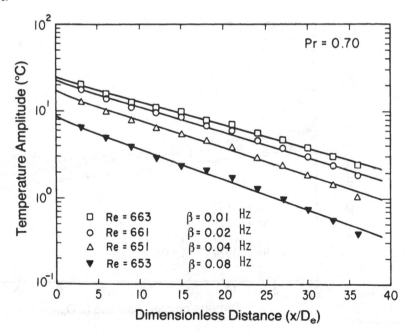

Fig. 14.4 Decay of temperature amplitude along the duct in laminar flow for $Re \approx 970$ and $Pr \approx 0.70$.

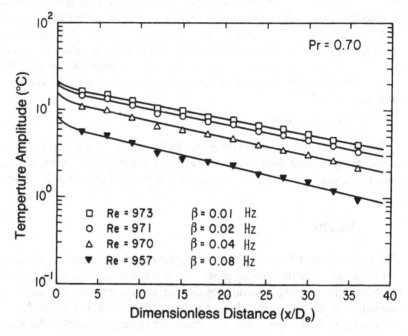

Fig. 14.5. Decay of temperature amplitude along the duct in turbulent flow for $Re \approx 11{,}000$ and $Pr \approx 0.70$.

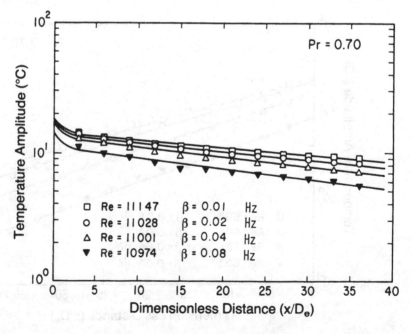

Table 14.1. *Experimental decay indexes α for laminar flow*

Re	β	α	Re	β	α	Re	β	α
491.16	0.01	0.0891	663.71	0.01	0.0616	843.48	0.01	0.0484
486.34	0.02	0.0917	661.25	0.02	0.0667	842.34	0.02	0.0514
487.17	0.04	0.0953	651.68	0.04	0.0722	831.96	0.04	0.0552
488.38	0.08	0.1006	653.03	0.08	0.0807	860.23	0.08	0.0581
973.58	0.01	0.0415	1102.39	0.01	0.0362	1396.00	0.01	0.0271
971.00	0.02	0.0435	1101.35	0.02	0.0389	1396.00	0.02	0.0295
970.16	0.04	0.0473	1100.76	0.04	0.0409	1396.00	0.04	0.0322
957.31	0.08	0.0517	1088.81	0.08	0.0457	1396.00	0.08	0.0364

Table 14.2. *Experimental decay indexes α for turbulent flow*

Re	β	α	Re	β	α	Re	β	α
5774.00	0.01	0.0102	9042.00	0.01	0.0080	11146.99	0.01	0.0126
5774.00	0.02	0.0178	9042.00	0.02	0.0101	11028.44	0.02	0.0153
5774.00	0.04	0.0157	9042.00	0.04	0.0126	11000.75	0.04	0.0173
5774.00	0.08	0.0196	9042.00	0.08	0.0174	10973.65	0.08	0.0193
15902.09	0.01	0.0111	18983.04	0.01	0.0110	20103.73	0.01	0.0105
16070.63	0.02	0.0142	18938.99	0.02	0.0137	20078.75	0.02	0.0132
16128.51	0.04	0.0165	18926.98	0.04	0.0157	20192.80	0.04	0.0157
16094.83	0.08	0.0193	18922.46	0.08	0.0183	20480.50	0.08	0.0178

Fig. 14.6. Decay of temperature amplitude along the duct in turbulent flow for $Re \approx 16,000$ and $Pr \approx 0.70$.

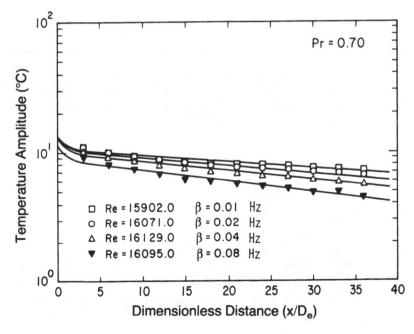

except at points very close to the entrance. For a given value of inlet frequency β, the value of the temperature amplitude at a point downstream depends on the Reynolds number Re. The higher the Reynolds number Re, the slower the decrease of the temperature amplitude along the duct, that is, the smaller value of the decay index α. For a given value of Re, the value of the temperature amplitude at a point downstream also depends on the inlet frequency β; when frequency β increases, the decay of the temperature amplitude will increase.

Suggested headings

Inlet Frequency: $\beta = $ ___ Hz;   *Geometry*: $b = $ ___ m; $L = $ ___ m;
$D_e = $ ___ m;   *Working fluid*: $\rho = $ ___ kg/m^3; $\nu = $ ___ m^2/s;
$m = $ ___ kg/s; $Re = $ ___ .

Temperature:

x/D_e	0 (Inlet)	1	2	3	....	α
T_{max}						....
T_{min}						....
ΔT						

References

1. ASME Standard, "Measurement of fluid flow in pipes using orifice, nozzle and venturi," MFC-3M-1984, 1984.
2. Kakaç, S.; Li, W., and Cotta, R. M., "Unsteady laminar forced convection with periodic variation of inlet temperature," *Trans. ASME J. Heat Transfer* 112 (1990): 913–20.
3. Li, W., "Experimental and theoretical investigation of unsteady forced convection in ducts," Ph. D. Dissertation, University of Miami, Florida, 1990.

Notation

a	thermal diffusivity, m^2/s
a^*	fluid-to-wall thermal capacity ratio
b	half height of the duct, m
Bi	modified Biot number, $h_e d/k$
C	constant for orifice plate in Eq. (14.10)
C_n	constants in Eq. (14.6)
D_e	hydraulic diameter, $4Lb/(L + 2b)$, m
g_n	function in Eq. (14.7)
h_e	equivalent heat transfer coefficient combined external convection and conduction, W/m^{2}K

L	width of the duct, m
m	mass flow rate, kg/s
N_n	norm of eignfunction $Y_n(\eta)$, $\int_0^1 U(\eta)Y_n^2(\eta)d\eta$
$U(\eta)$	dimensionless velocity profile
x	axial coordinate, m
y	normal coordinate, m
$Y_n(\eta)$	eigenfunction for Eqs. (14.1)–(14.4)

Greek Symbols:

α	decay index
β	inlet frequency, Hz
λ	eigenvalue corresponding to Eqs. (14.1)–(14.4)
ρ	density, kg/m^3
$\theta(\xi,\eta,\tau)$	dimensionless temperature
Ω	dimensionless inlet frequency, $2\pi\beta b^2/a$
τ	dimensionless time, at/b^2
ξ	dimensionless axial coordinate, $x/(U_m b^2/a)$
η	dimensionless normal coordinate, y/b
ΔP	pressure drop across orifice plate, Pa
$\Delta T(y)$	inlet temperature amplitude profile
ΔT_c	temperature amplitude at the center of the inlet
$\Delta\theta(\eta)$	dimensionless inlet temperature amplitude profile, $\Delta T(y)/\Delta T_c$

Weigong Li

Dr. Weigong Li, a visiting assistant professor at Florida International University, received his Ph.D. from the Department of Mechanical Engineering, University of Miami, in December 1990. In addition to unsteady forced convection, he also researchs viscous fluid flow and heat transfer inside helicoidal pipes, and heat and mass transfer within porous media during microwave decontamination and decommissioning of radioactively contaminated concrete.

Sadik Kakaç

Dr. Sadik Kakaç is professor and chairman of the Department of Mechanical Engineering, University of Miami. He received his Ph.D. in the field of heat transfer from Victoria University of Manchester, UK. He represented Turkey abroad in various scientific organizations between 1965 to 1980. He joined the University of Miami in 1980. He was the recipient of the Alexander von Humboldt Research Award for Senior U.S. Scientist in 1989. For the past 30 years, Dr. Kakaç has focused his research efforts on steady and transient forced convection in single-phase flow and two-phase flow instabilities. He is the author and/or co-author of five textbooks, and has published more than 100 research papers as well as 13 edited volumes in the field of thermal science.

EXPERIMENT 15

Measurement of heat and mass transfer from a body in air–water mist flow

Contributed by
T. AIHARA *and* T. OHARA

Principle

By suspending a small quantity of water droplets in a gas stream, convective cooling of heated bodies is remarkably improved in comparison with single-phase gas cooling.

Object

This experiment demonstrates how convective heat transfer can be enhanced due to the evaporation of a thin water film, maintained on a heated surface by continuous impingement of water droplets, and the sensible-heat cooling by droplet impingement. The enhancement of heat transfer is governed by the temperatures of the air–water mist and a heated body, air-stream velocity, water-to-air mass flow ratio, size and spatial distribution of water droplets, wet-area fraction of the body surface, and so on.

Apparatus

The outlines of the air–water mist flow tunnels to be used for the heat-transfer experiments are illustrated in Figs. 15.1 and 15.2. Figure 15.1 shows a test tunnel of the open type for horizontal air–water mist flow;[5] Fig. 15.2 shows a test tunnel of the circulation type for vertical air–water mist flow.[2] In the test tunnel shown in Fig. 15.1, pressurized water is sprayed through ten hollow-cone spray nozzles which are arranged in a circle around the centerline of the spray chamber. The additional use of a flat spray nozzle is recommended to humidify and saturate the carrier air. It is better to locate these nozzles sufficiently upstream so that drops larger than the desired ones may settle out of the mist in the low-velocity spray chamber.

A heated test body is installed horizontally in the potential core of a two-dimensional jet flow in the test chamber. A typical example of a test body with uniform wall heat flux is illustrated in Fig. 15.3. As for a heated cylinder

99

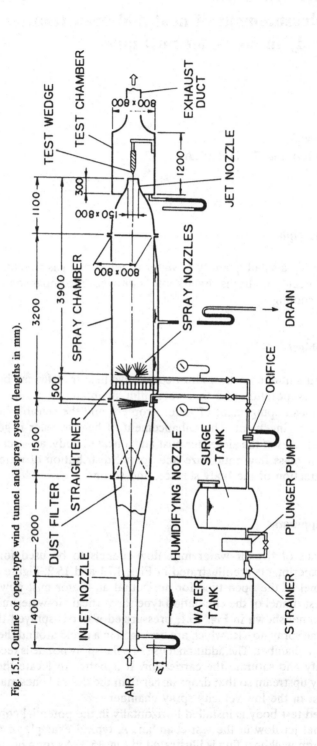

Fig. 15.1. Outline of open-type wind tunnel and spray system (lengths in mm).

Fig. 15.2. Outline of circulation-type wind tunnel (not to scale; lengths in mm).

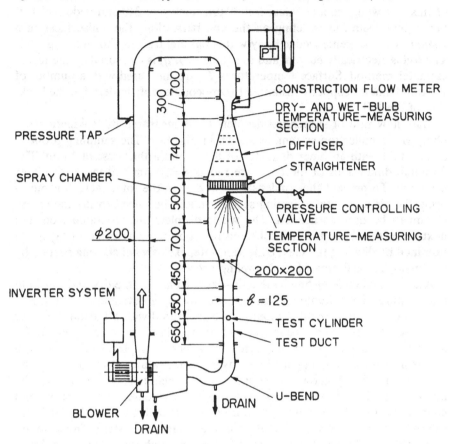

Fig. 15.3. Sectional view of test wedge and system of coordinates.

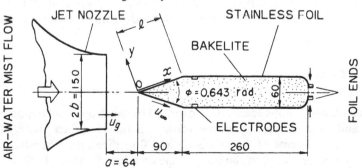

with a uniform wall temperature, refer to Aihara et al.'s work.[2] The surface of this test wedge is a stainless foil 30 μm thick and 240 mm wide which is fayed, by tensile forces acting on the end bars, along the midsurfaces of a 703 mm-wide Bakelite wedge body with a parallel trail, as shown in Fig. 15.3. The foil is electrically heated, and the Joule heat q_e is measured by the three-ammeter method. Surface temperatures T_w are measured with a number of 100 μm-diameter copper-constantan thermocouples soft-soldered to the back-side of the foil.

Mass flow rate of suspended droplets G_p is measured by isokinetic sampling with a collection system, as shown in Fig. 15.4. The sampling probe is a glass tube with an inner diameter of 8 mm and a thickness of 1 mm. The optimum dimensions of the bend are radius $r = 10$ mm and inside diameter $a = 5$ mm. To prevent the condensation of dew in the rotor meter for measuring the air flow rate, a heating section should be joined to the meter upstream to be used as needed. The water droplets are separated from the mixture in the separator and collected in a burette. The value of G_p is determined by dividing the mass collection rate, exclusive of starting period, by the cross-sectional area of the sampling probe.

After complete demisting by the air–water separator, the dry-and wet-bulb temperature of the humid air can be measured by a device as shown in Fig. 15.5. First, the air dry-bulb temperature is measured with a 100 μm-diameter copper-constantan thermocouple stretched in a glass tube. Subsequently, the air wet-bulb temperature is measured by a thermocouple of the same type that is imbedded in a copper block covered with saturated gauze, when the average air velocity across the copper block is maintained at 4 or 5 m/s by monitoring with an orifice flow meter. The thermometers, separator, and connecting line of the polyvinyl chloride tube should be fully thermally insulated to prevent dew condensation in the measuring system. The absolute humidity χ of the air–water mixture flowing through the test duct is determined from the measured dry- and wet-bulb temperatures.

In the test tunnel for vertical air–water mist flow (Fig. 15.2), the discharge air flows into the spray chamber of 0.5 m × 0.5 m cross section, through a dry-and wet-bulb temperature measuring section, a diffuser with five screens, and a straightener. City water whose temperature is controlled by a water-supply system is passed through the pressure controlling valve and a temperature measuring section; then it is sprayed through hollow-cone spray nozzles in a spray chamber and mixed with the air to produce an air–water mist mixture.

Procedure

The heat-transfer enhancement is governed by the droplet collection by a test body and then by the liquid-film formation on its surface, both of which depend on the shape of the test body and flow direction. Therefore, first of all, it is necessary to select the body shape and flow direction. In the following sections, experimental procedures are described for the case of a uniform heat-flux wedge in a horizontal air–water mist flow, as shown in Fig. 15.3.

Preliminary experiment with single-phase air flow

It is strongly recommended that a heat-transfer experiment with single-phase air flow be performed in order to check the accuracy of your experimental facilities/systems. The data on single-phase heat transfer are also required for comparison with those on the air–water mist test. If the measured heat-transfer coefficients agree satisfactorily with existing experimental data or established theoretical results, then you may proceed to the experiment with air–water mist flow.

Fig. 15.4. Collection probe for measuring mass flow rate of water droplets (not to scale).

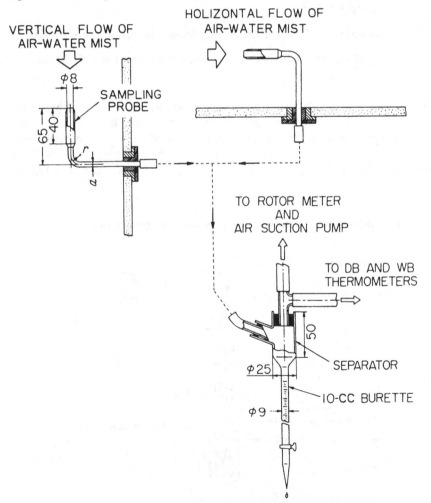

Production of air–water mist flow

Regulate the air flow rate to the desired value; then, spray and suspend water droplets in the air stream. Measure the size distribution of the droplets by an optical method or an immersion-sampling method.[4] A drop size of approximately 100 or 150 μm in mean diameter is recommended. The smaller the droplet, the more the droplet is liable to bypass the body; however, in the case of a much larger droplet, there is the possibility of a simultaneous increase in reentrainment loss due to liquid splashback from the liquid film.[1] Do not stop spraying water during the heat-transfer test to avoid the burnout of the test body. If the relative humidity of air is kept at 90 percent or higher, you may omit it from the group of parameters to be varied. The water-to-air mass flow ratio M is determined from Eq. (15.1):

$$M = \frac{G_p v_\infty}{u_g(1 + \chi_\infty)} \tag{15.1}$$

Measurement of heat-transfer coefficient for air–water mist flow

The local heat-transfer coefficient h (single-or two-phase) is defined as

$$h = q/(T_w - T_\infty) \tag{15.2}$$

where T_∞ is the dry-bulb temperature of humid air. The local wall heat flux q is approximately equal to the Joule heat q_e by electric heating. However, if you need precise measurements within an error of a few percent, you have to correct for the conduction effect.[5] Furthermore, the errors in T_w due to direct electric heating should be corrected by calibration. The vertical thin plastic

Fig. 15.5. Dry- and wet-bulb thermometers for air–water mists (not to scale).

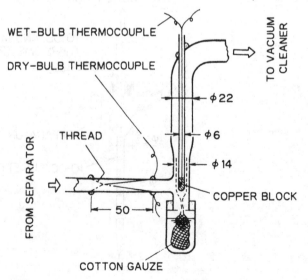

plates that are attached to both sides of the test body prevent the water film from escaping and keep the flow two-dimensional.

Results

Single-phase heat transfer from a uniform heat-flux wedge

Levy[6] proposed the following approximate expression of the single-phase heat-transfer coefficient $h_{(1)}$ for a laminar wedge flow of $Pr = 0.7$.

$$Nu_{x(1)} = 0.424 Re_x^{0.5} \tag{15.3}$$

where Nu_x and Re_x are the local Nusselt number and Reynolds number, defined as

$$Nu_x = hx/\lambda_g \tag{15.4}$$

$$Re_x = u_{g\infty} x/\nu_g \tag{15.5}$$

However, the free-stream turbulence in the jet potential core due to the spray system in the upstream will produce some increase in laminar heat transfer. Hence, Aihara et al.[5] derived the following empirical formula.

$$Nu_{x(1)} = 0.424 Re_x^{0.5} (0.94 + 1.04 \times 10^{-3} Re_x^{0.5}) \tag{15.6}$$

This formula correlates with their experimental data fairly well in the range $2 \times 10^3 < Re_x < 6 \times 10^4$.

Heat transfer from a uniform heat-flux wedge in an air–water mist flow

The following simplifying assumptions are introduced:

1. The presence of water droplets in the air stream does not affect the air flow around the wedge.
2. The droplets are uniformly distributed in the jet flow; they travel in straight paths with the mass velocity G_p and have the same temperature as the air wet-bulb temperature T'_∞.
3. On impinging upon the wedge surface, the droplets immediately heat up to the bulk temperature of the water film.
4. The heat transfer from the film to the air follows the empirical formula Eq. (15.6) for a dry wedge, and Lewis's relation holds between heat and mass transfer.
5. No reentrainment occurs.
6. The temperature drop across the water film is negligible.
7. The enthalpy transport by the water film is also negligible.

With the foregoing assumptions and from the time-averaged balances of mass and heat, the following approximate expression is derived for the fully wet region of the wedge surface.[5]

$$h_{(2)} \cong h_{(1)} \left[1 + \frac{(\hat{\chi}_w - \chi_\infty) r_w}{(T_w - T_\infty) c_g} \right] + \frac{T_w - T'_\infty}{T_w - T_\infty} G_p c_1 \sin\frac{\phi}{2} \qquad (15.7)$$

where ϕ is the apex angle of the wedge. In the dryout region where the droplets impinging on the wedge surface evaporate, the time-averaged balances of mass and heat are approximated as:[5]

$$h_{(2)} \cong h_{(1)} + \frac{r_w + c_1(T_w - T'_\infty)}{T_w - T_\infty} G_p \sin\frac{\phi}{2}. \qquad (15.8)$$

The typical measured values of the time-averaged, two-phase heat-transfer coefficient $h_{(2)}$ are shown in Fig. 15.6, where ΔT_{max} denotes the difference between the air dry-bulb temperature T_∞ and the maximum wedge surface temperatures, and the curves of $h_{(1)}$ for single-phase flow are evaluated from Eq. (15.6) for the same jet conditions as those of the two-phase runs. The addition of small quantities of water droplets to the air stream enhances the two-phase heat-transfer coefficients to 2–14 times the corresponding single-phase values. Whereas at high mass flow ratios the local heat-transfer coefficient $h_{(2)}$ increases with increasing ΔT_{max}, at low mass flow ratios it decreases with increasing ΔT_{max}. The latter tendency is particularly remarkable in the dryout regions, which are represented by the symbols $\triangle$ and $\triangle$. As a general tendency, the value of $h_{x(2)}$ increases as M, or G_p, and the jet velocity, or approaching velocity u_g, are increased. These experimental findings coincide well with the enhancement tendency of heat transfer resulting from Eqs. (15.7) and (15.8).

Now we consider the ratio of the measured heat-transfer coefficient to the value predicted according to Eq. (15.7), that is,

$$f_x = [h_{(2)}]_{exp}/[h_{(2)}]_{th}. \qquad (15.9)$$

Figure 15.7 is a plot of the average $\bar{f}$ with respect to the wedge surface length l, exclusive of the dryout data. It is shown by the experimental finding of $\bar{f} \simeq 1$ that Eqs. (15.8) and (15.9) give a good approximation for $h_{x(2)}$. Specifically, at relatively low mass flow ratios, $\bar{f}$ is greater than unity owing to assumption 4. However, with increasing mass flow ratio, assumptions 3 and 6 become invalid due to the thickening water film and consequently $\bar{f}$ tends to decrease gradually; after reaching a minimum at $M \simeq 2 \times 10^{-2}$, $\bar{f}$ begins to increase again. This re-increase in $\bar{f}$ may be attributed to an increase in turbulence and to assumption 5. On the other hand, the smaller the mass flow ratio, the higher is the rate of droplets that diminish in size with evaporation while passing through the thermal boundary layer and flow downstream without impinging on the wedge surface. This is the reason why at $M = 4.7 \times 10^{-4}$ the values of $\bar{f}$ decrease with increasing ΔT_{max}.

Heat transfer from a cylinder with uniform wall temperature in an air–water mist flow

Figure 15.8[1] is a typical plot of the local Nusselt number $Nu_{d(2)}$, defined by Eq. (15.10) for a horizontal circular cylinder in a downward flow with a relative humidity of 98 percent or more for most runs, using the vertical tunnel, as shown in Fig. 15.2.

$$Nu_d = hd/\lambda_g \qquad (15.10)$$

$$Re_d = u_{g\infty}d/v_g \qquad (15.11)$$

In this case, a continuous thin water film is generally formed on the surface of the front half of the test cylinder owing to droplet impingement. The physical properties of the humid air are to be evaluated at the film temperature and film humidity $\chi_f = (\chi_w + \chi_\infty)/2$. The numerical solutions of Aihara, Fu, and Suzuki,[3] indicated by the solid lines in Fig. 15.8, are in very good agreement with the present experimental data.

There are considerable differences in the distribution of the local Nusselt number in the rear half of the cylinder between the clear air flow and the air–water mist flow. In the separated region of clear air flow, a reverse flow creeps up from the backward stagnation point of the cylinder toward the separation point; as a consequence, the local Nusselt number $Nu_{d(1)}$ increases greatly with β. In contrast to this, the local Nusselt number $Nu_{d(2)}$ for the air–water mist flow takes a minimum value at $\beta \simeq 2\pi/3$ in the case of a large temperature difference, whereas it gradually increases with β in the case of a small temperature difference. The reasons for this are that the water film

Fig. 15.8. Comparison of local nusselt number $Nu_{d(2)}$ for air–water mist flow between the experimental data[2] and the numerical solutions[3] for $Re_d = (1.38 - 1.62) \times 10^4$ and $d/b = 0.4$; $d_0 = 142\,\mu m$ and $n = 3.7$ for curves 1 and 2; $d_0 = 108\,\mu m$ and $n = 3.4$ for curve 3.

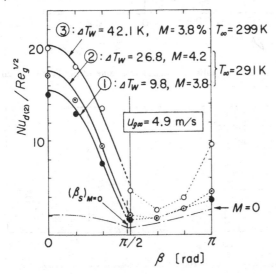

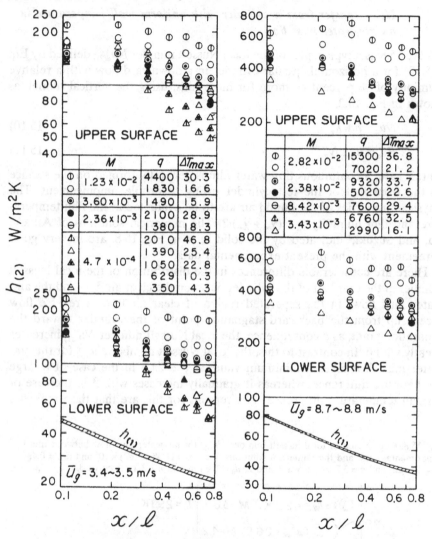

Fig. 15.6. Local heat-transfer coefficients for air–water mist flow $h_{(2)}$. The relative humidity for the majority of data is above 88 percent.[5]

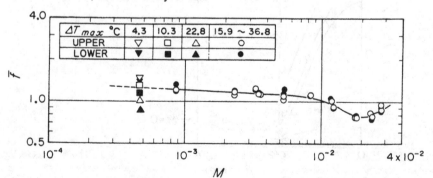

Fig. 15.7. Relation between the average ratio $\bar{f}$ of measured value to predicted one and the mass flow ratio M, exclusive of dryout data.[5]

on the cylinder has the antipodal characteristics of evaporative cooling and thermal resistance and the wet-area fraction ξ varies according to the governing parameters Λ and Φ defined by Eqs. (15.12) and (15.13).

$$\Lambda = MPrRe_d^{1/2}\,(c_l/c_g) \tag{15.12}$$

$$\Phi_w = \frac{(\hat{\chi}_w - \chi_g)r_w}{(T_w - T_\infty)c_g} \tag{15.13}$$

The local enhancement factors of heat transfer, $Nu_{d(2)}/Nu_{d(1)}$, are plotted in Fig. 15.9, where the heat-transfer mechanisms can be seen more clearly.

Suggested headings

X [m]	T_w [K]	T_∞ [K]	T'_∞ [K]	q [W/m²]	u_g [m/s]	G_p [kg/(m²·s)]

In the case of Fig. 15.3, u_g is the average velocity at the nozzle mouth.

References

1. Aihara, T., "Augmentation of convective heat transfer by gas–liquid mist," *Proc. 9th Int. Heat Transfer Conf., Jerusalem* 1 (1990): 445–61.

Fig. 15.9. Effect of temperature difference ΔT_w ($= T_w - T_\infty$) on distribution of local enhancement factor $Nu_{d(2)}/Nu_{d(1)}$ for $Re_d = (0.8 - 2.2) \times 10^4$ and $d/b = 0.4$; $d_0 = 168\,\mu m$ and $n = 3.7$ for curve 4; $d_0 = 111\,\mu m$ and $n = 3.4$ for curve 5; $d_0 = 120\,\mu m$ and $n = 3.7$ for curve 6; $d_0 = 168\,\mu m$ and $n = 3.7$ for curve 7; other parameters and symbols are the same as in Fig. 15.8.[2]

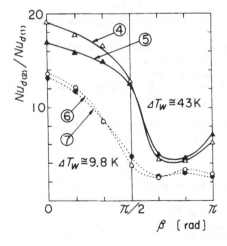

2. Aihara, T.; Fu, W.-S.; Hongoh, M., and Shimoyama, T., "Experimental study of heat and mass transfer from a horizontal cylinder in downward air–water mist flow with blockage effect," *Experimental Thermal and Fluid Science* 3 (1990): 623–31.
3. Aihara, T.; Fu, W.-S., and Suzuki, Y., "Numerical analysis of heat and mass transfer from horizontal cylinders in downward flow of air–water mist," *J. Heat Transfer* 112 (1990): 472–8.
4. Aihara, T.; Shimoyama, T.; Hongoh, M., and Fujinawa, K., "Instrumentation and error sources for the measurement of the local drop-size distribution by an immersion-sampling cell," *Proc. 3rd Int Conf. Liquid Atomization and Spray Systems, London* 2 (1985): VC/5/1–11.
5. Aihara, T.; Taga, M., and Haraguchi, T., "Heat transfer from a uniform heat flux wedge in air–water mist flows," *Int. J. Heat Mass Transfer* 22 (1979): 51–60.
6. Levy, S., "Heat transfer to constant-property laminar boundary-layer flows with power-function free-stream velocity and wall-temperature," *J. Aeronaut. Sci.* 5 (1952): 341–8.

Notation

b	m	tunnel width
c	kJ/(kg·K)	specific heat at constant pressure
d	m	diameter of circular cylinder
d_0	μm	size parameter in Rosin–Rammler equation
G_p	kg/(m²·s)	mass flow rate of suspended droplets
h	W/(m²·K)	local heat-transfer coefficient
l	m	length of wedge surface
M		water-to-air mass flow ratio defined by Eq. (15.1)
n		dimensionless dispersion parameter in Rosin–Rammler equation
Nu_x, Nu_d		local Nusselt number, defined by Eqs. (15.4) and (15.10), respectively
Pr		Prandtl number
q	W/m²	local wall heat flux
r	kJ/kg	latent heat of evaporation
Re_d, Re_x		Reynolds number, defined by Eqs. (15.5) and (15.11), respectively.
Re_g		gas Reynolds number, $d_c u_c/\nu_g$
T_∞, T'_∞, T_w	K	temperatures of dry bulb, wet bulb, and local wall, respectively
u_g, u_∞	m/s	approaching air velocity and air velocity at the edge of boundary layer, respectively
v	m³/kg	specific volume of humid air
x, y	m	Cartesian coordinates, see Fig. 15.3
β	rad	azimuth angle from the forward stagnation point
ΔT_w	K	$T_w - T_g$
Λ		coolant-feed parameter, defined by Eq. (15.12)
λ_g	W/(m·K)	thermal conductivity of air
ν_g	m²/s	kinematic viscosity of air
ξ		wet-area fraction, defined by the percentage of total heat-transfer area covered by a liquid film
ρ	kg/m³	density
Φ		evaporation parameter, defined by Eq. (15.13)
χ	kg/kg	absolute humidity of humid air

Subscripts:

g	gas phase or humid air
l	water
w	heated wall or at T_w
(1)	single-phase (air) flow

(2)   air–water mist flow
∞   far upstream or free stream

Superscripts:

‾   average
^   saturated value

Toshio Aihara

Toshio Aihara received his Doctor of Engineering
degree in heat transfer from Tohoku University in
1968. Dr. Aihara has been associated since 1966 with
the Institute of Fluid Science (former Institute of
High Speed Mechanics), Tohoku University, as a
professor. He has published more than 87 papers
and articles and 9 books in the basic and applied
research areas of gas–liquid mist cooling, fluidized-
bed heat exchangers, thermal stability of
superconductors, heat transfer of supercritical fluids,
free-convection heat transfer from various bodies
and ducts, compact heat exchangers, and heat-
transfer control.

Taku Ohara

Taku Ohara received his D. Eng. degree from the
University of Tokyo in 1991. He is currently
associated with the Institute of Fluid Science,
Tohoku University, as a research associate. His
research is in the areas of fluid dynamics and heat
and mass transfer such as liquid-film flow, solid-
particle behavior in liquid flow, gas–liquid two-phase
flow, shock-wave propagation in bubbly liquid, and
thin-film formation in spin coating.

EXPERIMENT 16
Measurement of transient/steady heat-transfer coefficient with simultaneous photography of flow processes from beneath the heater surface

Contributed by
HERMAN MERTE, JR.

Objective

A thin metallic film, with appropriate properties and firmly attached to an insulating substrate, can perform simultaneously as a heater and resistance thermometer precisely located at the surface of the substrate. This provides a well-defined boundary condition from which the transient heat-transfer rates and the heat-transfer coefficients to the fluid can be determined. If the film is thin enough to be transparent the fluid flow can be visualized with tracer particles, or boiling behavior can be observed without interference from intervening bubbles. The heater surface can be as large or small as desired.

Figure 16.1 presents the concept using a gold film deposited on a quartz substrate. The heater can be used in several ways. If the fluid remains motionless, the temperature distribution in both the quartz substrate and the fluid can be computed from the measured power input, using classical techniques, and the measured surface temperature can be compared with the computed surface temperature. Once the fluid has been set in motion, whether

Fig. 16.1. Concept of transient heating with a thin film.

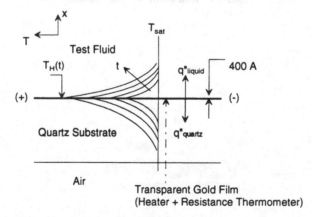

112

by natural or forced convection or by boiling, the transient measurements of the thin-gold-film temperature and the power input permit the computation of the mean heat flux to or from the substrate, and hence to the fluid. A limitation of the temperature measurement with this heater surface is that only the integrated mean surface temperature is measured. With forced convection over the surface and a uniform imposed heat flux it can be anticipated that a temperature variation will arise in the flow direction as the thermal boundary layer develops. For steady-state operation the fraction of heat input transferred to the fluid with forced convection and/or boiling can be determined once the steady heat loss through the substrate is known as a function of the interfacial and the surrounding-substrate-holder temperatures. This is obtained by calibration.

Apparatus

One scheme for rendering the concept of Fig. 16.1 into an operational device is shown in Fig. 16.2, and it shows the means by which the current-carrying and the potential lead electrical connections are carried through the quartz surface without introducing any impediments to the fluid flow. The assembly can be installed so that the heater surface is flush with the container walls.

A 400-angstrom-thick gold film is sputtered on the quartz substrate, which was presputtered with a 30-angstrom tantalum layer for improved gold adhesion. Before the tantalum layer is applied, the surface is cleaned by ion bombardment to remove about 10 angstroms of the quartz substrate. The thin film serves simultaneously as a heater and a resistance thermometer, and has a negligible time constant associated with the transient temperature measurement. The temperature of the thin film thus is identical to that of the surface of the quartz substrate, which is polished with a 1.4-micron pitch polish prior to the coating process. The semitransparent gold surface illustrated has a rectangular shape 1.19 cm by 3.81 cm (0.75 in. by 1.5 in.), and is considerably larger than the size of the bubbles formed above it, for the nucleate boiling applications. This semitransparent gold surface has a higher resistance than the 10,000-angstrom-thick gold surface near the power taps, which act as the current connection. The voltage-measuring leads also consist of 10,000-angstrom-thick gold films. Kovar screws passing through holes drilled in the quartz are used as power and voltage taps. The low linear thermal expansion coefficient of Kovar closely matches the linear thermal expansion coefficient of the quartz substrate, which is mounted in a Teflon holder. This circular holder can be rotated in its own plane, so that the heated length or aspect ratio in the flow direction with forced convection can be changed conveniently by a factor or two. This will have the effect of changing the thermal-boundary-layer thickness for given levels of flow velocity and heat flux. Since the working fluid is in direct contact with the gold film, it is necessary that non-electrical-conducting fluids be used.

With an appropriately designed vessel, the processes occurring at the heater surface can be viewed simultaneously from the side and through the heating surface. With boiling this permits the obtaining of data on the departure size and trajectory of the bubbles, along with the nucleation site density and frequency of bubble departures.

As in the work of Oker,[4,5] it is found that the electrical resistance of the gold surfaces varies with time, but that the slope of the linear relationship of surface resistance and temperature, dR/dT, for each particular surface remains constant. This resistance change is attributed to an aging effect. As per

Fig. 16.2. Gold-film heater-surface assembly.

Gold Film Surface

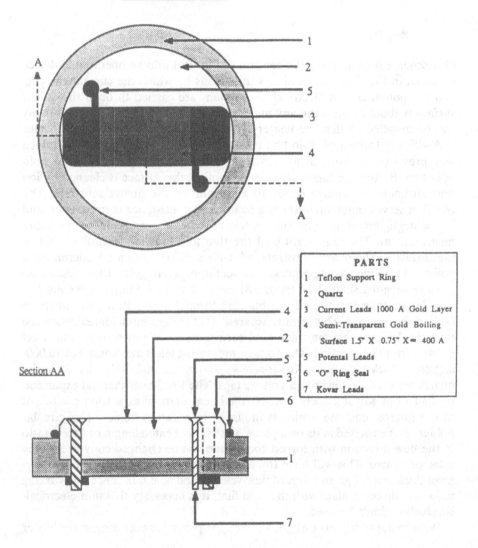

PARTS	
1	Teflon Support Ring
2	Quartz
3	Current Leads 1000 A Gold Layer
4	Semi-Transparent Gold Boiling
	Surface 1.5" X 0.75" X ≈ 400 A
5	Potential Leads
6	"O" Ring Seal
7	Kovar Leads

Section AA

the suggestion of Oker,[4] the surfaces are tempered at a temperature of about 275 °C for 20 minutes, which accelerates the aging to a stable level. It has also been found in early testing that sputtered gold surfaces are more durable than those deposited by vacuum evaporation. The temperature–electrical-resistance relation for each thin gold film surface are determined by calibration. It has been established that for the gold films used, appropriate calibrations and voltage measurements to 5 significant figures are needed with reasonable precautions to obtain accuracies of about ±0.6 °C (±1 °F) for the mean surface temperature. These precautions include taking an "in situ" calibration immediately preceding each experiment.

Procedure

Transient boiling experiments are performed using step increases in the imposed heater surface heat flux. The transient heater surface temperature for a representative test with pool boiling is shown in Fig. 16.3, along with a tabulation of the test conditions. This applies to the case where the heater surface is facing up in the Earth's gravity. The onset of natural convection appears as an irregularity in the temperature–time plot. The time from the energization of the heater to the onset of natural convection is designated as t_{nc} and is identified by the departure of the heater surface temperature from the one-dimensional semi-infinite media transient conduction solution, and also by the observed onset of fluid motion as characterized by a wavelike disturbance recorded photographically. The next event following the onset of natural convection in Fig. 16.3 is incipient boiling, as indicated.

Up to the onset of natural convection, the heat transfer is diffusional, with the transient interfacial temperature for the step input heat flux at the interface of two semi-infinite solids given by Carslaw and Jaeger (1959) as:

$$\frac{T_i(t) - T_o}{q''T} = \frac{2(\alpha_Q \alpha_1 t)^{1/2}}{\pi^{1/2}(k_Q \alpha_1^{1/2} + k_1 \alpha_Q^{1/2})} \tag{16.1}$$

Eq. (16.1) is plotted in Fig. 16.3 as the 1D transient conduction prediction. An analytical expression for the transient heat-transfer coefficient for the 1D conduction domain to the fluid is obtained by dividing the imposed heat flux to the fluid by the instantaneous difference between the heater surface temperature and the initial bulk fluid temperature, with the result:

$$h_1 = \frac{k_1 \pi^{1/2}}{2(\alpha_1 t)^{1/2}} \tag{16.2}$$

It is to be noted that the heat-transfer coefficient defined in this way for the transient 1D conduction domain is independent of the level of the heat flux imposed, and is a function only of the fluid properties. This is plotted in Fig. 16.4 for R-113.

Using the measured surface temperature in Fig. 16.3 as an input, the transient temperature distribution in the quartz substrate can be determined with a finite-difference computational scheme. From a polynomial fit of the discrete temperatures near the heater surface the heat flux to the substrate can be computed, from which the mean heat flux to the fluid is determined, knowing the power input flux to the thin gold film. With the mean heat flux to the fluid, an appropriately defined heat-transfer coefficient can be calculated. The technique described previously thus provides a means for measuring transient heat-transfer coefficients, and steady-state ones as well, from which it is possible to determine when steady-state conditions have been reached.

Fig. 16.3. Measured transient heater surface temperature for horizontal surface. Working fluid: R-113. $q_T^{\prime\prime} = 5.7$ W/cm^2.

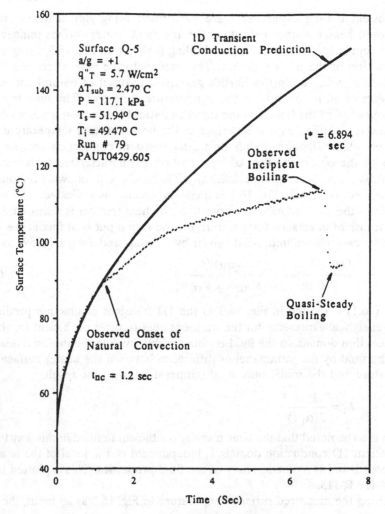

Results

Figure 16.5 presents the temperature distributions computed from the meas-ured surface temperatures for the transient diffusional process in both the quartz and R-113, at one-second time intervals. The temperature distribu-tion shown in the R-113 is not at all realistic after one-second because of the natural convection then present. Figure 16.6 shows the computed heat flux into the quartz substrate. It becomes negative upon nucleation, indicating that it is releasing heat to the R-113. The average heat flux to the R-113 is the difference between the measured power input and that computed to the quartz substrate. The heat-transfer coefficient can be determined from this heat flux and the temperature difference between the heater surface and

Fig. 16.4. Transient R-113 heat-transfer coefficient for 1 D-conduction heat-transfer domain.

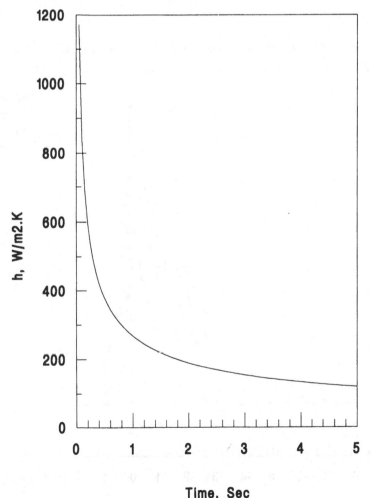

the bulk liquid, and is plotted in Fig. 16.7. A comparison is shown with the predicted heat-transfer coefficient using a standard steady-state correlation, with the details given in the sample calculation. The early part of the "measured" heat-transfer coefficient in Fig. 16.7 compares reasonably well, up to about 1 second, with the predicted conduction domain given in Fig. 16.4. The undershoot and subsequent overshoot are consequences of transient natural-convection liquid momentum effects.

Heater surface-temperature data for an earlier, somewhat noisy, test at a higher heat flux level are shown in Fig. 16.8, with the corresponding heat-transfer measurements computed from those given in Fig. 16.9. It may be noted that almost steady values for both natural convection and boiling are reached rather quickly.

Some typical photographic results of transient boiling obtained with a heater surface such as in Fig. 16.2 are presented by Ervin and Merte[1] and Ervin et al.[2]

Fig. 16.5. Computed temperatures, from measured surface temperatures of Fig. 16.3.

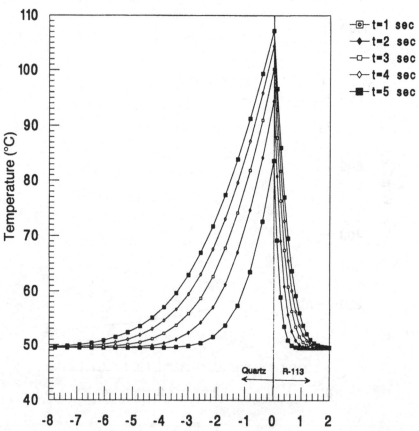

Transient voltage measurements necessary to produce the transient mean heater surface temperatures, such as the examples given in Figs. 16.3 and 16.8, are conveniently made with any of a variety of high-precision microprocessor-controlled data acquisition systems available on the commercial market. The computations and plotting such as Figs. 16.5–16.7 likewise can be automated with readily available computing equipment.

Sample calculation

Heat flux applied: $5.7 \ W/cm^2$
$T_\infty = 49.5 \ ^\circ C \ (322K)$: Bulk temperature
$T_s = 110 \ ^\circ C \ (383K)$ at $t = 6$ sec: surface temperature
$T_f = (49.5 + 110)/2 \ ^\circ C = 352.9K$: Film temperature

Fig. 16.6. Substrate heat flux determined from measured surface temperatures of Fig. 16.3.

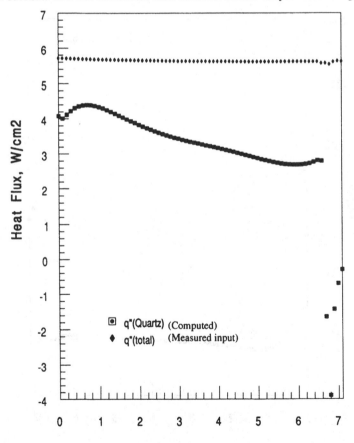

Time, Sec

Properties of R-113 at T_f:
$\alpha = 0.434 \times 10^{-7}$ m²/s: thermal diffusivity
$k = 63.87 \times 10^{-3}$ W/m · K: conductivity
$v = 25.61 \times 10^{-8}$ m²/s: kinematic viscosity
$g = 9.8$ m/s²: gravity
$\rho = 1425.0$ kg/m³
$\rho_{383} = 1340$ kg/m³
$\rho_{322} = 1504.0$ kg/m³
$\mu = 365 \times 10^{-6}$ Pa · s

Using the natural convection correlation of Lloyd and Moran,[3] with the corresponding characteristic length,

$$Nu_L = 0.15 Ra_L^{1/3} \text{ for } Ra_L > 8 \times 10^6$$

Fig. 16.7. Transient natural-convection heat-transfer coefficient leading to nucleation. Data of Fig. 16.3.

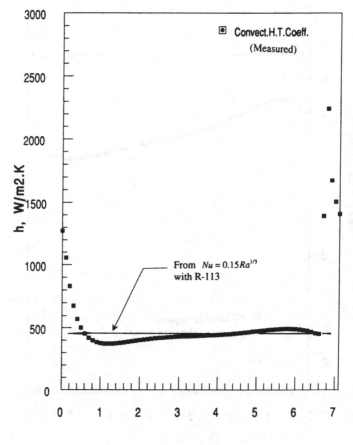

where

$$Ra_L = \frac{g\beta(T_s - T_\infty)L^3}{\alpha v}$$

$$Nu_L = hL/k$$

$$L = As/P = w\tfrac{1}{2}(w+1) = 38.1 \times 19.5 \times 10 - 6/2(38.1 + 19.5)10^{-3}$$
$$= 0.00645 \text{ m}$$

$$\beta = -\frac{1}{\rho}\frac{\partial\rho}{\partial T} \cong -\frac{1}{\rho}\frac{\rho_\infty - \rho}{T_\infty - T} = -\frac{1}{1340}\frac{1504.2 - 1340.0}{322.65 - 383.15}$$

$$= 2.0254 \times 10^{-3} \text{ K}^{-1}$$

$$Ra_L = \frac{g\beta(T_s - T_\infty)L^3}{\alpha v}$$

$$= \frac{9.8 \times 2.0254 \times 10^{-3} \times (110 - 49.5) \times 0.00645^3}{0.434 \times 10^{-7} \times 25.6 \times 10^{-8}} = 2.898 \times 10^7$$

Fig. 16.8. Measured transient heater surface temperature for horizontal surface. Working fluid: R-113. $q_T'' = 7.5$ W/cm².

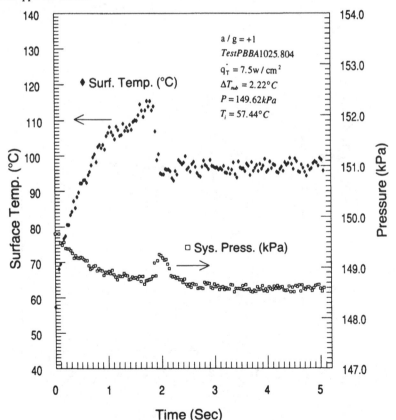

$$Nu_L = 0.15 \ Ra_L^{1/3} = 46.067$$

$$h = Nu_L \ k/L = 46.077 \times 63.87 \times 10 - 3/0.00645 = 456 \ \text{W/m}^2 \cdot \text{K}$$

References

1. Ervin, J. S., and Merte, H., Jr., "A fundamental study of nucleate pool boiling under microgravity," Final Report on NASA Grant NAG3-663, UM Report UM-MEAM-91-08, Department of Mechanical Engineering and Applied Mechanics, University of Michigan, Ann Arbor, MI, 1991.
2. Ervin, J. S.; Merte, H., Jr.; Keller, R. B., and Kirk, K., "Transient pool boiling in microgravity," *Int. J. Heat Mass Transfer* 35, 3 (1992): 659–74.
3. Lloyd, J. R., and Moran, W. R., "Natural convection adjacent to horizontal surface of various planforms," *Trans. ASME J. Heat Transfer* 96C, 4 (1974): 443–7.

Fig. 16.9. Transient natural-convection heat-transfer coefficient leading to nucleation. Data of Fig. 16.8.

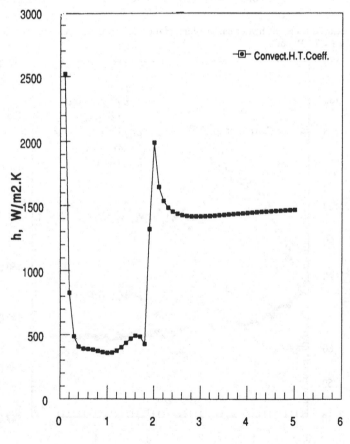

4. Oker, E., and Merte, H., Jr., "Transient boiling heat transfer in saturated liquid nitrogen and F113 at standard and zero gravity," U. Michigan, Rept. No. 074610-52-F for NASA, 1973.
5. Oker, E., and Merte, H., Jr., "Semi-transparent gold film as simultaneous surface heater and resistance thermometer for nucleate boiling studies," *J. Heat Transfer* 103 (1981): 65–8.

Herman Merte, Jr.

Herman Merte, Jr., is Professor of Mechanical Engineering at the University of Michigan, where he has been a member of the faculty since 1960. He was a visiting professor at the Technical University of Munich in 1974–5.

PART I.3
Boiling

Fig. I.3.1. Film boiling occurs at temperatures higher than those which cause transition boiling. Here the tube is surrounded by a transparent film of alcohol vapor. At the top of the copper tube the vapor has formed into a wavy "rod" about to break into a row of bubbles. (Photographed by J. W. Westwater, *Sci. Am.* 190, 6 (1954): 64–8.)

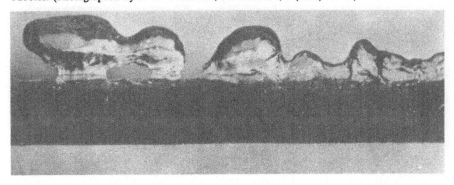

Fig. I.3.2. Transition boiling occurs at temperatures higher than those causing nucleate boiling. Note the slugs of alcohol vapor forming explosively, masking the copper tube. Also note the slug of alcohol vapor surging away from the tube at lower right. (Photographed by J. W. Westwater, *Sci. Am.* 190, 6 (1954): 64–8.)

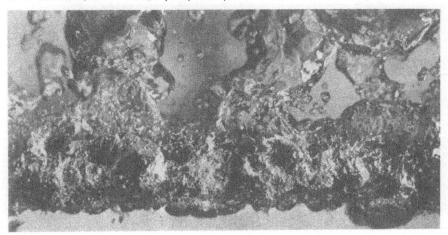

Fig. I.3.3. Nucleate boiling photographed at 1×10^{-6} second in a glass tank filled with methyl alcohol. Visible is a horizontal copper tube heated by steam. The bubbles of alcohol form repeatedly at the same points on the tube surface. (Photographed by J. W. Westwater, *Sci. Am.* 190, 6 (1954): 64–8.)

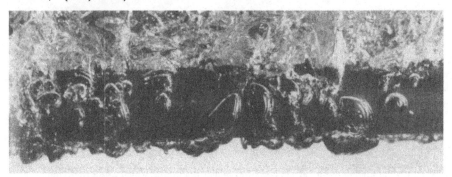

EXPERIMENT 17
Capillary-tube boiling

Contributed by
MIHIR SEN

Principle

The process of nucleate boiling from a heated surface involves bubble formation, bubble emission, and liquid replacement in a cyclic manner. The phenomenon is strictly periodic at low heat fluxes, becoming gradually aperiodic
as the heat flux increases.

Object

This is an experiment to provide some understanding of the complex phenomena that occur during liquid-to-vapor phase change. The geometry is
greatly simplified so that the process can be easily controlled and most of the
results visually observed. Periodicity and aperiodicity of the bubbling can be
quantitatively analyzed through a study of the bubble departure periods.

Apparatus

Figure 17.1 (not to scale) shows the arrangement used in the experiment. The
capillary tube in which boiling is to be studied is constructed in the following
manner. A thin electrical heater wire (of constantan, diameter 75 μm, for
instance) is uniformly roughened with emery paper to discourage preferential
nucleation at any particular spot on its surface. It is then run down the center
of a 1–3-mm-diameter glass capillary; a 4–7-cm length of this capillary tube
is closed off by heating over a flame. A DC power source supplies variable
current to the electrical wire. The power from this source can be determined
by measuring its voltage and amperage. The length of the wire should also be
determined to provide the heat flux in W/m of the heater length.

   The tube with its heater wire is fixed in a vertical position in a beaker filled
with water. The beaker is placed on a flat-plate electric heater. Temperature

All figures have been taken or adapted from Acharya[1] or Acharya and Sen.[2]

127

fluctuations at the lip of the capillary, induced by the passage of a vapor–liquid interface during bubbling, are sensed by a very fine (75 μm has worked well) differential copper-constantan thermocouple. The two junctions of the thermocouple are placed at points A and B so that the voltage produced at the thermocouple terminals is roughly proportional to the difference in temperature between those locations. The differential arrangement removes most of the effect of temperature fluctuations due to the motion of water outside the capillary arising due to convection. The thermocouple is intended only to mark the instant of bubble passage, and hence need not be calibrated. The time constant τ of the thermocouple, previously estimated by dipping it suddenly into hot water and observing its response on an oscilloscope, should not be more than about 30 ms. This is fast enough for the frequencies observed in the present experiments. The thermocouple signal is amplified and fed into a data acquisition system, and then stored in a computer data file for further processing.

If available, a video camera with a macro lens can also be used for a close-up visual record of the boiling process. The frames can be analyzed in slow motion to reveal the physical processes which take place during boiling. This procedure also permits measurement of the bubble diameter and the amplitude and frequency of the lower interface motion.

Procedure

All surfaces are cleaned with acetone and then with distilled water prior to use. Distilled water is also used as the working fluid; degassing is achieved by

Fig. 17.1. Schematic of capillary tube boiling.

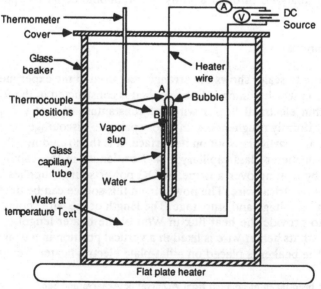

boiling the water for about 30 minutes before commencing the test runs. The flat-plate heater is used to keep the temperature in the beaker T_{ext} at about 2 °C below saturation. The heat flux to the capillary-wire heater is varied by changing the current passing through it.

At very small heat fluxes, there is no boiling observed as all the input heat is transferred away to the water in the beaker. At larger heat fluxes, boiling is observed near the top of the capillary. A vapor slug is formed, the upper and lower interfaces of which move in different directions. The lower interface moves down pushing liquid up along the walls of the capillary. The upper interface grows outside the capillary and forms an almost spherical bubble. Finally when the bubble breaks off liquid moves in to fill the void, and the cycle begins again. Bubble emission becomes quite complicated for high heat fluxes. Care must be taken not to reach film boiling, at which point the heating wire becomes blanketed by an insulating vapor film leading to overheating and damage to the wire.

The bubbles affect the thermocouple reading as they move past the junction. The instants of vapor bubble passage correspond to spikes on a thermocouple voltage–time trace stored in the computer. A simple program can later read the data stored in the computer and find the time interval between successive bubbles ($t_1, t_2, \ldots$).

Results

Measurement of average frequency

Figure 17.2 shows part of a typical temperature trace from the thermocouple at fairly low heating for which the bubble formation is seen to be approximately periodic. The average frequency over N samples is $\bar{f} = 1/\bar{t}$, where $\bar{t} = (\Sigma t_i)/N$.

As the heat flux increases, the bubbling will become less periodic. However, the average frequency can still be determined in the same manner. The data can then be plotted as a frequency-versus-heat-flux curve. A typical

Fig. 17.2. Temperature trace for periodic boiling.

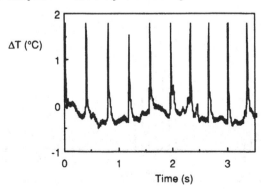

example is shown in Fig. 17.3. In this case boiling begins around 20 W/m, and the heat flux increases up to 80 W/m.

Statistics of aperiodic boiling

At high heat fluxes boiling is not periodic, and a quantitative statistical analysis of the periods can be made. From the periods t_1, t_2, ... a histogram can be constructed, an example of which is shown in Fig. 17.4. As the heat flux increases, the width of the histogram also increases, indicating that the phenomenon becomes less periodic. Other quantities such as the variance or standard deviation can also be determined to provide similar information.

Fig. 17.3. Bubbling frequency for different heat fluxes.

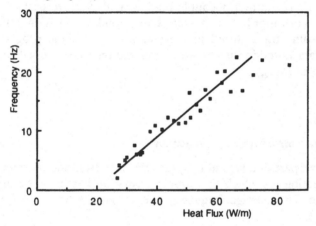

Fig. 17.4. Histogram of bubbling period.

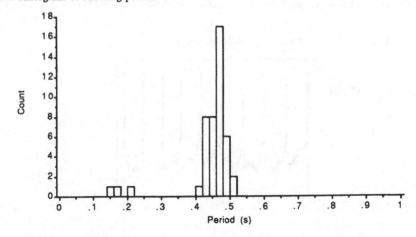

Period-doubling phenomenon

At low heat fluxes the boiling phenomenon is periodic, as shown by the temperature trace in Fig. 17.2. At a certain heat flux period doubling occurs, after which the boiling cycle is composed of a small period followed by a larger one. An example of the period-two state is shown by the temperature trace in Fig. 17.5. This phenomenon is seen more clearly by plotting a return map as follows. From a series of periods t_1, t_2, ..., plot t_n vs. t_{n+1} for $n = 1$, 2, ... A periodic signal will give a small region where most of the points fall as in Fig. 17.6, whereas a doubly periodic signal will show two such regions as in Fig. 17.7.

Fig. 17.5. Temperature trace for doubly periodic boiling.

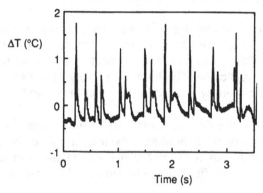

Fig. 17.6. Return map for periodic boiling.

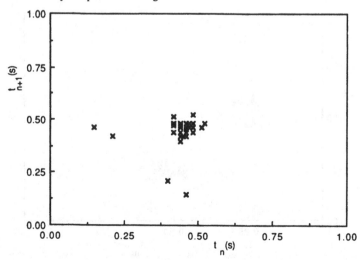

Explanation

Boiling from a flat surface is normally a very complex phenomenon. Detailed studies can be found in books such as Collier[4] and Carey.[3] In our case the process has been simplified greatly by use of a relatively simple capillary geometry. A complete cycle consists of bubble formation, growth, emission, and filling of the void. Some hot liquid is pushed out along the walls of the capillary during bubble growth. Cooler liquid also comes in during the void-filling phase of the cycle. This means that a major part of the heat being supplied by the heater goes not to the vapor but to the water outside the capillary.

We can, however, assume that a fixed fraction of the input heat goes into the vapor that is formed. In addition if we take the bubble diameter to be proportional to the capillary diameter, we have the heat balance relation

$$(\dot{Q} - \dot{Q}_o)L \sim \bar{f}\rho_{vapor}\frac{4\pi R^3}{3}h_{fg}$$

where $\dot{Q}_o$ is the heat flux necessary for boiling to start, taking into account the heat loss from the capillary. This relation implies that the frequency and the heat flux are linear. This is true only for low heat fluxes. After that the process becomes more complicated. The limits of validity of the expression can be experimentally explored by using different heat fluxes and capillary-tube geometries.

Period doubling is due to the dynamic instability of the boiling cycle. For further information on this or other nonlinear dynamical effects the reader can consult Sen[5] or many recent texts on the matter. As the heat flux increases, the speed with which the fluid moves in and out of the capillary tube

Fig. 17.7. Return map for doubly periodic boiling.

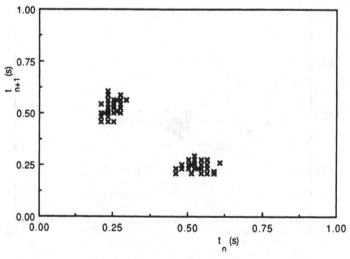

also increases. In addition, a larger vapor slug is formed than can be incorporated into a single bubble. For these reasons, two bubbles of different sizes are observed for each boiling cycle. This process gets very complicated at high heat fluxes, at which stage the periods between successive bubbles appear to vary randomly. This will show up in the t_n-versus-t_{n+1} plots as well as in statistical analysis of the period data.

Suggested headings

Constants:

$T_{ext} =$ _____; $\tau =$ _____

$I =$ _____; $V =$ _____

n	t_n
1	
2	
3	
4	

References

1. Acharya, N., "Deterministic chaos in nucleate boiling," M.S. Thesis, Department of Aerospace and Mechanical Engineering, University of Notre Dame, Notre Dame, Indiana, 1989.
2. Acharya, N., and Sen, M., "Frequency measurements in capillary tube boiling," *Proceedings of the 2nd International Multiphase Fluid Transient Symposium*, ed. M. J. Braun, FED-vol. 87, pp. 13–19, ASME, New York, 1989.
3. Carey, V. P., *Liquid–Vapor Phase Change Phenomena*, Series in Chemical and Mechanical Engineering, Hemisphere Publishing, Washington, DC, 1992.
4. Collier, J. G., *Convective Boiling and Condensation*, McGraw-Hill, New York, 1972.
5. Sen, M., "The influence of developments in dynamical systems theory on experimental fluid mechanics," *Frontiers in Experimental Fluid Mechanics*, Lecture Notes in Engineering 46, ed. M. Gad-el-Hak, Springer-Verlag, Berlin, 1989.

Notation

$\bar{f}$	average bubble frequency
h_{fg}	latent heat of vaporization
I	heater current
L	capillary tube length

N	number of samples
$\dot{Q}$	heat flux per unit length
$\dot{Q}_o$	heat flux necessary for boiling to start
R	capillary tube radius
$\bar{t}$	average bubble period
t_i	period between two successive bubbles
V	heater voltage
ρ_{vapor}	density of vapor

Mihir Sen

Mihir Sen did his graduate studies at the Johns Hopkins University and the Massachusetts Institute of Technology. After that he worked at the National University of Mexico in Mexico City for ten years. He was a visiting professor at Cornell University before taking up his present faculty position in the Department of Aerospace and Mechanical Engineering of the University of Notre Dame. His research interests include heat transfer and fluid mechanics.

EXPERIMENT 18

Two characteristic regions of nucleate pool boiling and corresponding change of hydrodynamic state

Contributed by
Y. KATTO

Principle

Existence of two different characteristic regions in nucleate boiling, that is, the region of isolated bubbles at low heat fluxes and the region of interference at high heat fluxes, can be demonstrated by measuring the resultant force acting on a compact heater that is suspended in a pool of liquid with boiling on its upper surface.

Object

As the surface temperature of a heater submerged in a pool of saturated liquid is raised above the saturation temperature, nucleate boiling appears after the incipience of boiling, and then continues up to the point of critical heat flux. In this nucleate-boiling regime, the heat flux from the heated surface to the liquid increases with increasing surface temperature, and the relationship between the heat flux and the surface temperature (the so-called boiling curve) is of a monotonic nature, in most cases exhibiting no noticeable change of character throughout the aforementioned regime. Visual observations by means of a high-speed cine camera, however, reveal that nucleate boiling is divided into two regions. Namely, when the heat flux is low, small isolated bubbles repeat the growth and departure process at a comparatively small number of active nucleation sites on the heated surface, but at high heat fluxes, the heated surface is covered with a thin liquid sublayer (which is generally called the macrolayer[1]) holding numerous tiny vapor jets rooted to nucleation sites, and the vapor fed continuously from these vapor jets accumulates to develop massive vapor slugs successively on the foregoing liquid sublayer. This transition of flow configuration between the two different regions is very important, because it leads to the state exhibiting the phenomenon of critical heat flux, that is, the upper limit of heat flux in nucleate boiling; and the foregoing two regions can be readily discriminated by measuring the resultant force acting on the boiling surface.

135

Apparatus

Liquid vessel
Cylindrical compact heater
Transparent inner wall (glass plate)
Fluid-motion preventer
Auxiliary heater
Condenser
Liquid (water; distilled water if possible)
Balance
Mercury junction
Variable auto-transformer (for cylindrical and auxiliary heaters)
Digital multimeter to measure electric power

A version of the experimental apparatus is shown in Fig. 18.1, which consists
of three main parts – a liquid vessel, a cylindrical heater, and a balance. The
cylindrical heater is suspended by an arm of the balance so that the resultant
force acting on the heater surface can be measured. The liquid vessel is
divided by means of transparent inner walls into the inner and the outer
spaces. The cylindrical heater is then suspended in the inner space, while an
auxiliary heater is set up in the outer space in order to keep the liquid in the
vessel at saturation temperature. A fluid-motion preventer, which is made of
a thin metal plate with a hole at the center to allow the vertical movement
of the cylindrical heater, is employed to suppress the motion of liquid in the
space under the preventer plate. The inner space, which is constructed so as
to have a hexagonal (or cylindrical) cross section, is about 150 mm across and
300 mm in height. In order to prevent the change of the liquid level with time
in the vessel, it is advisable to use a simple condenser (not shown in Fig. 18.1)
that can return the condensate to the vessel. The cylindrical heater of diameter
$D = 45$ mm and height $H = 40$ mm typically is enclosed by a stainless-steel

Fig. 18.1. Main portion of experimental apparatus (after Katto and Kikuchi[2]).

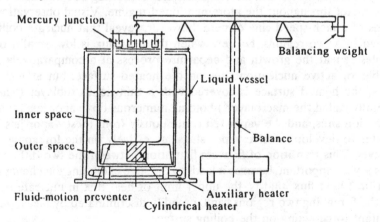

wall with a cross section such as that illustrated in Fig. 18.2. The top end of a copper block stored in the heater plays the role of a heated disk surface of about 15 mm in diameter. A thin stainless-steel plate (preferably less than 0.5 mm in thickness to prevent heat loss) is brazed to the circumference of the heated disk surface. The lower part of the copper block, about 30 mm in diameter, has many grooves to store electric heating elements. The space between the copper block and the heater wall is filled with an adequate heat-insulating material. The leakage of heat to the parts other than the heated surface is generally small. (If correct values of the heat flux q at the heated surface are desired, three Almel-Chromel thermocouples are set up along the axis of the upper part of the copper block to measure the temperature gradient, through which q can be estimated by knowing the magnitude of thermal conductivity of the copper block.) The arm of the balance is specially designed so as to fit the size of the liquid vessel. The cylindrical heater is suspended from the arm of the balance about 200 mm to the left of the fulcrum, while an adjustable balancing weight is put on the right end of the arm. A dish for weights is suspended at each position 100 mm left and right of the fulcrum. Mercury junctions are employed for the supply of electric power to the heater, and for the measurement of temperature by the foregoing thermocouples if it is wanted.

Procedure

Fill the vessel with hot water up to a level L above the upper surface of the cylindrical heater (the depth of water), which may be chosen between 20 and 200 mm.

Fig. 18.2. Cross section of cylindrical compact heater (after Katto and Kikuchi[2]).

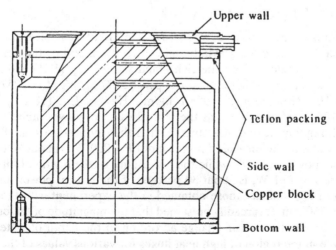

Upper wall

Teflon packing

Side wall

Copper block

Bottom wall

Heat the water in the vessel rapidly by the auxiliary heater until boiling
    begins on the heated surface. Then reduce the electric power to a
    low value capable of sustaining very weak boiling.
Adjust the balancing weight as well as the weights on the dishes so as to
    make the arm of the balance horizontal; this is the standard position
    to measure the buoyant force acting on the heater under boiling.
Start the supply of electric power to the cylindrical heater, and increase the
    power input step-by-step until several active nucleation sites appear
    on the heated disk surface.
Fix the magnitude of the power input P and wait several minutes to obtain
    a steady state of heat conduction in the copper block. Then, measure
    the increase of the buoyant force B caused by the foregoing boiling
    relative to the standard state mentioned before. In order to avoid the
    error inevitable from the fluctuation of boiling, it is advisable to
    repeat the same measurement a few times.
Increase the power input P step-by-step and repeat the foregoing sequence
    at each step.

Explanation

As for the buoyant force acting on the heater caused by boiling, the measurement will reveal such characteristics as shown in Fig. 18.3, where the
increase of buoyant force B (in grams) is plotted against the power input P
(in Watts). In this case, the transition between the two characteristic regions
of nucleate boiling is observed to appear at $P = 75.1$ W for the heated surface of diameter $d = 14.5$ mm. Hence, if heat loss is neglected for simplicity,
it means that the transition appears at heat flux $q = P/(\pi d^2/4) = 4.55 \times 10^5$
W/m^2. Meanwhile, for the magnitude of critical heat flux q_c, there is the well-
known Kutateladze correlation:

$$(q_c/\rho_v H_{fg})/[\sigma g(\rho_L - \rho_v)/\rho_v^2]^{1/4} = k \qquad (18.1)$$

where ρ_v denotes the density of vapor, H_{fg} the latent heat of evaporation,
σ the surface tension, g the gravitational acceleration, and ρ_L the density of
liquid. The magnitude of k on the right-hand side of Eq. (18.1) is 0.18. For
saturated water at atmospheric pressure, $\rho_v = 0.5977$ kg/m^3, $H_{fg} = 2257$ kJ/kg,
$\sigma = 58.92 \times 10^{-3}$ N/m, and $\rho_L = 958.1$ kg/m^3. Hence, Eq. (18.1) predicts
$q_c = 1.52 \times 10^6$ W/m^2, which means the critical value of power input $P_c = 251$
W for the foregoing heated disk surface. It is then noticed in Fig. 18.3 that
the highest value of the measured power input $P = 200$ W is about 80 percent
of the critical power input P_c, while the lower limit of the region of high heat
fluxes, where $P = 75.1$ W, is about 30 percent of the critical power input P_c.
    The data in Fig. 18.3 are those obtained in the experiment with the depth
of water $L = 150$ mm. It is readily presumed that the magnitude of the buoyant
force acting on the heater can change according to the depth of water. The
buoyant force in the region of high heat fluxes for various values of the depth

of water L in the range 50 to 200 mm is illustrated in Fig. 18.4, where the ordinate represents the increase in the buoyant force divided by the area of the upper surface of the cylindrical heater of diameter $D = 42$ mm, that is, the reduction of the mean pressure on the upper surface of the heater, $\Delta p = B/(\pi D^2/4)$. The abscissa of Fig. 18.4 represents the heat flux across the heated surface of diameter $d = 14.5$ mm.

It is noticed from Figs. 18.3 and 18.4 that when the depth of water L is kept constant, the mean pressure on the upper surface decreases with increasing heat flux. Very roughly speaking, this may be related to the increase of vapor volume existing between the heated surface and the free surface of water. However, there is a peculiar phenomenon observed in Fig. 18.4 that, as L

Fig. 18.3. Increase of buoyant force acting on heater due to boiling for diameter of heated surface $d = 14.5$ mm and depth of water $L = 150$ mm (after Katto and Kikuchi[2]).

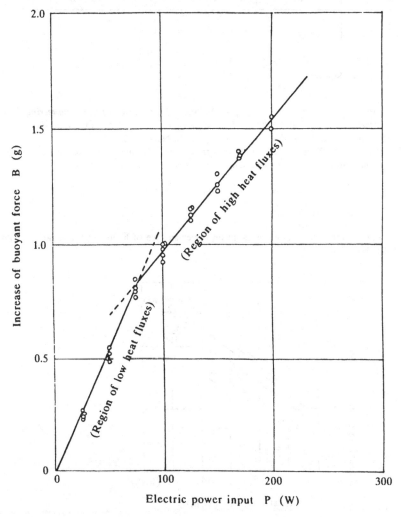

increases above 50 mm, the mean pressure on the upper surface of the heater decreases up to a minimum value at $L = 150$ mm, and thereafter the mean pressure begins to increase with L (see the data for $L = 150, 175$, and 200 mm). This change of trend suggests that when the depth of water is large, the natural convection in the liquid vessel may noticeably change the pressure distribution in the field of interest.

Suggested headings

Constants: $d =$ _____; $D =$ _____; $L =$ _____

P(W)	B(g)	q(W/m²)	Δp(mmAq)	Observation

References

1. Katto, Y., "Critical heat flux in pool boiling," *Proceedings of the Engineering Foundation Conference on Pool and External Flow Boiling*, eds. V. K. Dhir and A. E. Bergles, pp. 151–64, American Society of Mechanical Engineers, New York, 1992.
2. Katto, Y., and Kikuchi, K., "Study of forces acting on a heated surface in nucleate boiling at high heat fluxes," *Heat Transfer-Japanese Research* 1, 4 (1972): 36–46.

Fig. 18.4. Reduction of mean pressure on upper surface of heater in the region of high heat fluxes (after Katto and Kikuchi[2]).

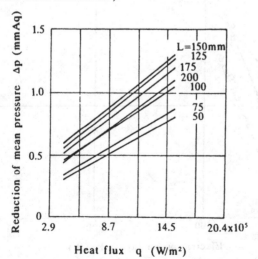

Y. Katto

Y. Katto received a D.Eng. degree (1960) from the University of Tokyo. He joined the Institute of Science and Technology of the University of Tokyo in 1947, moved to the National Aerospace Laboratory in 1956, and became a professor of heat transfer at the Department of Mechanical Engineering of the University of Tokyo in 1963. At the present time, he is a professor of heat transfer at the Department of Mechanical Engineering of Nihon University, Tokyo.

EXPERIMENT 19
The boiling slide

Contributed by
PETER GRIFFITH

Principle

An image of an electrically heated wire boiling under water is projected on a screen. Boiling in real time is easily observed.

Object

This experiment provides a greatly enlarged, easily controlled demonstration of boiling. The image is so clear the audience leaves with an unforgettable picture of the processes that constitute boiling.

Apparatus

Power is brought into the projector through the posts that are fastened to the lid of the slide. (The lid is shown in profile in Fig. 19.1.) The posts are fastened to leads that are made of springy copper strips with clips on the ends which hold the wire from which the boiling occurs. When the whole device is assembled it looks like Fig. 19.1.

The wire is the only part of the assembly that is critical. We have found that Chromel A wire 0.010 inches in diameter is appropriate. This wire passes through critical heat flux (CHF, burnout, boiling transition, etc.) when an AC current of 6.4 amps passes through it. Very little more current than this also causes the wire to physically burn out.

The power for this wire is provided by a Variac which has a maximum output of 10 amps. An ammeter in the line passing the Variac output is needed to insure that the maximum allowable current is not exceeded.

For demonstrative purposes, the slide projector, Variac, and ammeter are mounted on a wheeled cart. Figure 19.2 is a view of the whole system assembled. All these components are hard wired to an extension cord for which an adapter for every kind of wall outlet should be provided. The cart and

Fig. 19.1. Boiling slide assembly with the cover removed for the plan view.

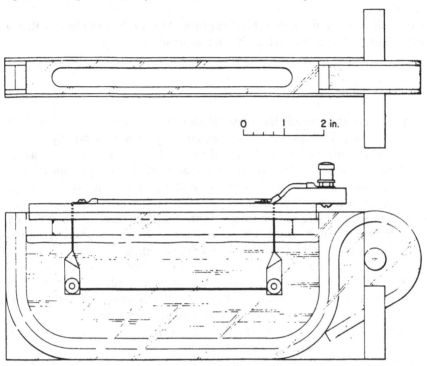

Fig. 19.2. Assembly for the boiling slide demonstration: 1. the slide (Fig. 19.1); 2. the ammeter (10A); 3. the variac (10A); 4. squeeze bottle with replacement water; 5. envelope with extra wires (the wires are 6″ long, 0.01″ in diameter, and made of chromel or stainless steel); 6. the slide projector with a large slot.

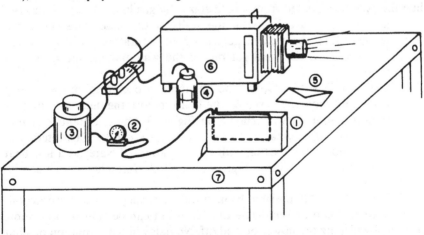

apparatus are wheeled to the class room and the view projected on the wall or a screen.

Extra wires, each about 6-inches long, are always taken to class so that a physical burnout does not cause the demonstration to fail.

Procedure

In order to use this device, the vessel is almost filled with water. The lid is put on and the power turned up. Heating up is most rapidly accomplished by starting at 10.0A and cutting back to 6A when the water is at saturation. The current must be turned down promptly when approaching saturation or the wire will burn out. It takes 12 to 20 minutes from a cold start to get up to saturation.

The boiling slide should not be inserted into the slide machine until shortly before it is to be used. The steam that it emits will condense on the lenses of the slide machine and spoil the image.

The image is projected inverted so rising bubbles appear to go down rather than up. We have tried a number of lens and mirror arrangements but have always found that the loss in the sharpness of the image is more annoying than the fact that bubbles rise down.

Condensation on the lenses is no problem when the slide projector is on because the projector runs pretty hot.

After about 20 minutes of boiling full tilt, replacement water must be added. For this purpose a small plastic squeeze bottle of clean water should be taken to class and added as needed. The cold water suppresses boiling for 3 or 4 minutes so don't add replacement water just before the start of the demonstration.

When the system is heated from a cold start, the gas in the water comes out of solution and collects on the walls of the slide. This spoils the image. Once the system gets close to saturation, however, these bubbles grow, rise, and get out of the way and the image is very clear.

We've found it is best to describe the phenomena occurring on the boiling curve and show the curve on the board before the demonstration. A typical boiling curve is shown on Fig. 19.3. Quite a number of features that affect what we see on this figure can be illustrated with the boiling slide. Starting with almost-saturated water at a low heat flux (at the lower left in Fig. 19.3) the following phenomena can be illustrated:

1. With 1 or 2 A passing through the wire, single-phase natural convection can be seen. The time constant can be shown to be short by flicking the power on and off. No delay in the formation of what appears to be a fully developed natural-convection boundary layer is evident. If one stays in this regime long enough the wire will become covered with air bubbles.

2. Before real boiling begins the surface of the air bubbles on the wire will flow due to the Marangoni effect. Jets of hot water will be projected in various directions from bubbles on the surface as the interface flows due to temperature differences between the base and the pole of the bubble.

3. An increase in the power to 3 A or so will cause boiling to start. Before there are too many bubbles on the wire the power can be flicked on and off a couple of times and the fact that bubbles originate at points attached to the wire can be shown. The residue of air rising from the condensed bubbles will be evident.

4. A further increase in the power will cause an increase in both the bubble population and the bubble-formation frequency at the existing active sites. At some point the bubbles will be so close together that it will be difficult to fit in any more. An increase in the population of bubbles could lead to a reduction in heat transfer. Above 6 A, increase the power about 0.10 A at a time. At some point around 6.4 to 6.8 A, CHF will occur and the look of the boiling will change dramatically.

5. When CHF occurs, a continuous tube of vapor surrounds the wire. By increasing or decreasing the power to the wire very slightly, the tube of vapor surrounding the wire can be made to slowly grow or decrease in length. The co-existence of the two dramatically different heat-transfer regimes at the same power level can be easily demonstrated.

6. By removing the slide from the projector, the fact that part of the wire is glowing because it is much hotter can be shown to the whole class.

Fig. 19.3. Typical boiling curve for water at one atmosphere of pressure. The details of the shape depend on the geometry of the surface.

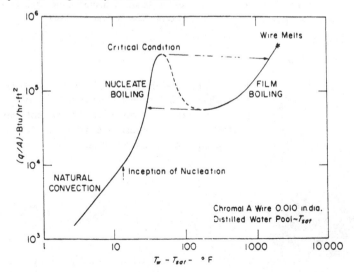

7. If the power is turned down a little bit, the film-boiling section length can be stabilized. Flicking the power off and then on again will cause the film boiling to collapse and nucleate boiling to be reestablished everywhere. In this way the hysteresis of the boiling curve can be shown.

8. If the power is set to about 0.1 A below the power where CHF occurred, and the slide removed from the projector, the effect of gravity on CHF can be illustrated. Hold the slide up high and "drop" it a couple of feet. (That is, suddenly lower it.) Usually a local transition to film boiling can be made to occur and that transition can be seen by the whole class because part of the wire is glowing. The film boiling region is usually stable and will persist even though the power is still below the value that causes a spontaneous transition to film boiling.

9. The changed bubble dynamics in film boiling are easy to show too. When part of the wire in film boiling and part in nucleate the regular release of large bubbles in the film-boiling region is clear. The alternate locations of the release of bubbles from the wire are clear too from the pattern of bubbles rising above the wire.

Conclusion

This is a simple, fail-safe demonstration that is remembered years after students who have taken heat transfer have forgotten most of what you told them. It is cheap, easy to build, thought provoking, and fun to operate.

Reference

1. Bergles, A. E., and Griffith, P., "Projection slides for classroom demonstration of heat transfer with boiling and condensation," *Bull. Mech. Enging. Educ.* 6 (1967): 79–83.

Peter Griffith

Peter Griffith received his Sc.D. at M.I.T. in 1956. Since then he has been on the faculty of M.I.T. in the Mechanical Engineering Department. His primary research interests are two-phase flow, heat transfer, and nuclear reactor safety.

EXPERIMENT 20

Evaporation and boiling in sessile drops on a heated surface

Contributed by
SHIGEAKI INADA *and* WEN-JEI YANG

Object

The objectives of this experiment are as follows:

(i) to observe the evaporation and boiling phenomena in a drop after it falls onto a heated surface,
(ii) to measure the lifetime of the drop, and
(iii) to determine the heat-transfer characteristics in the sessile drop-boiling system.

Apparatus

Figure 20.1 is a schematic of the experimental setup. It consists of a drop-generating system and an electrically heated testing surface with a thermal measurement device.

The liquids used in the drops are carbon tetrachloride, benzene, methyl alcohol, and distilled water.

A liquid fills a bottle equipped with a stopcock which is attached to a support and hangs over the heated surface. A needle valve regulates the rate of dripping flow from the bottle through a no.-1/4 hypodermic needle. The liquid is released from the needle in the form of drops 2 to 3 mm in diameter which fall at regular intervals.

In order to generate drops of identical size, a drop receiver is used to collect the drops until they fall at a steady, desired rate.

At the start of the experiment, the receiver is quickly displaced via the action of a spring to allow only a single drop to fall onto the center of the heated surface.

A center rest-pin is installed on the drop receiver to produce two-dimensional movement (in order to prevent drop-receiver vibration).

Blotting paper or tissue paper is hung at the exit of the drop receiver to smoothly drain the collected liquid.

147

The heated surface is one end of a cylindrical block that is slightly concave up in order to prevent a drop from jumping out when it rolls or splashes on the surface. It is made out of a material such as copper, stainless steel, brass, or aluminum. Nonoxidizing materials, such as platinum or carbon, or low

Fig. 20.1. Schematic of the experimental set up.

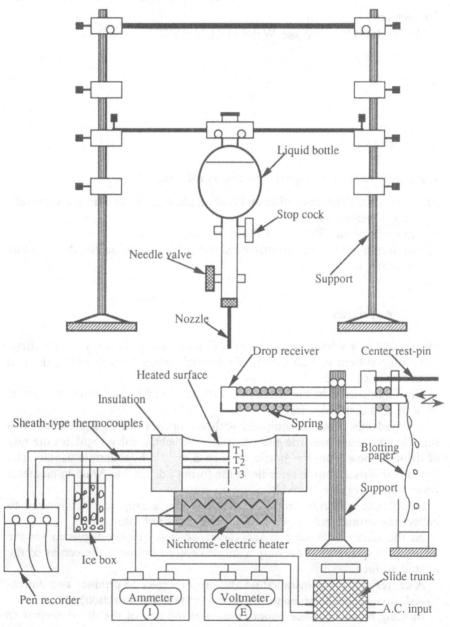

heat-conducting materials, such as ceramics, may also be used for the heated surface.

The cylindrical block is heated by radiation from a Nichrome electric heater which is placed underneath it. The heated-surface temperature is obtained by an extrapolation of temperatures measured by three sheath-type thermocouples that are installed at three locations on the centerline of the cylindrical block. Experiments are conducted under the atmospheric pressure with drops initially at room temperature.

Procedure

Adjust the slide trunk so that the input electric current and voltage are increased in steps.

In each step, after confirming that the temperature being monitored by the pen recorder has become steady, a single liquid drop is allowed to fall on the heated surface.

A stopwatch is employed to measure the time from the moment of contact between the drop and the heated surface until the end of drop evaporation, called drop lifetime.

During each drop evaporation, it is necessary to investigate both the drop behavior (situations of evaporation and boiling, drop bouncing, break-up, and drop rolling and spinning) as well as the boiling sound. Hence, a careful adjustment of the heated-surface temperatures is very crucial.

The heating process must go through each power increment, avoiding a sudden power increase.

Drop-lifetime curve

Three local temperatures in the cylindrical block are extrapolated to determine the heated-surface temperature T_w.

The drop lifetime τ (the time from the drop–surface contact to the disappearance of the drop) is obtained using a stopwatch.

τ (ordinate) is plotted against T_w (abscissa). The experiment is repeated for the same drop size, varying T_w from a low value (in the natural-convection regime) to a high value (in the spheroidal-evaporation, or film-boiling, region). The resulting curve, called the drop-lifetime curve, takes an inverted N-shape, as seen in Figs. 20.2 and 20.3.[2] It is just the opposite of a conventional N-shaped pool-boiling curve.[1,4] There is one minimum and one maximum point on the drop-lifetime curve. The minimum point is referred to as the maximum-boiling-rate point, beyond which a transition-type boiling phenomenon takes place within the drop. The maximum point is called the Leidenfrost point beyond which the drop takes a spherical form and is separated from the heated surface by a very thin vapor film.

One can superimpose a schematic of observed drop behavior and the magnitude of boiling sound directly on the lifetime curve.

Weber number for drop impact on a heated surface

The experimental device allows an adjustment in the height from which drops are released onto the heated surface. The impact velocity of a drop on the heated surface V_0 can be varied by changing the height h. The impact Weber number is defined as

$$We = \rho V_0^2 D_0 / \sigma \qquad (20.1)$$

Here, ρ denotes liquid density, V_0 impact velocity ($= \sqrt{2gh}$), D_0 drop initial diameter, and σ surface tension of the liquid. The Weber number is a dimensionless number whose magnitude decides whether or not a drop will break up upon impact with the heated surface.

Fig. 20.2. Lifetime curve for water, $D_0 = 1.88$ mm.[1]

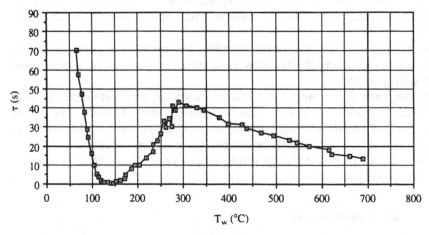

Fig. 20.3. Lifetime curve for benzene, $D_0 = 2.14$ mm.[1]

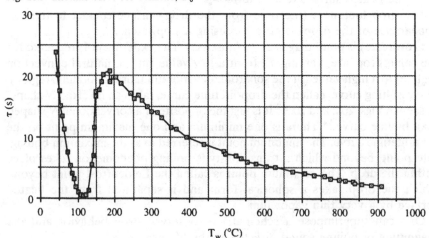

Changes in drop diameter during the evaporation process

In the film-boiling regime where T_w exceeds the Leidenfrost point, the instantaneous drop diameter $D(\tau)$ changes with time as

$$D^2(\tau) = D_0^2 - C\tau \tag{20.2}$$

Here, C is the evaporation-velocity coefficient,[3] which is a characteristic property of an individual liquid.

Analysis

Identify and mark the maximum-boiling-rate point and the Leidenfrost point on the drop-lifetime curve. Indicate the spheroidal regime in which a drop may roll on the heated surface. Investigate how the Weber number and drop materials affect those points and the regime.

Draw a typical pool-boiling curve and mark on it the maximum-heat-flux point (i.e., the burnout point) and the minimum film-boiling heat-flux point. Investigate how the boiling curve is related to the drop-lifetime curve obtained from the present experiment.

The naked eye can clearly observe whether or not a drop breaks up upon contact with the heated surface whose temperature exceeds the maximum-boiling-rate point. A break up of drops occurs when the Weber number exceeds a certain value. Determine the critical Weber number.

Determine the magnitude of the evaporation-velocity coefficient C in Eq. (20.2), when the heated-surface temperature is in the spheroidal-evaporation regime.

Apply the heat-transfer viewpoint to explain drop behavior in the spheroidal-evaporation regime. Explain the force balance to support a drop on a thin vapor film.

Describe the difference in boiling sounds at the maximum-boiling-rate point and in the transition regime (between the maximum-boiling-rate point and the Leidenfrost point). Investigate the cause of the difference.

Suggested headings

Kind of liquid _____

$\rho =$ _____ ; $\sigma =$ _____

$D_0 =$ _____ ; $V_0 =$ _____ ; $We =$ _____

lifetime τ	temperature of block				power to heater		evaporation-velocity coefficient	boiling sounds and drop behavior
	T_w	T_1	T_2	T_3	$I(A)$	$E(V)$	C	

References

1. Miyasaka, Y.; Inada, S., and Izumi, R., "Study of boiling-characteristic curves in subcooled pool-boiling of water," *International Chemical Engineering*, 23, 1 (1983): 48–55.
2. Tamura, Z., and Tanasawa, Y., "Evaporation and combustion of a drop contacting with a hot surface," *7th Symposium (International) on Combustion*, pp. 509–22, Butterworths Scientific Publications, London, 1959.
3. W.-J. Yang, "Vaporization and combustion of liquid drops on heated surfaces," *Two-Phase Transport and Reactor Safety*, eds. T. N. Veziroglu and S. Kakac, pp. 51–67, Hemisphere Publishing, Washington, DC, 1978.
4. Zhang, N., and Yang, W.-J., "Evaporation and explosion of liquid drops on a heated surface," *Experiments in Fluids* 1 (1983): 101–11.

Wen-Jei Yang

Wen-Jei Yang received his Ph.D. in 1960 from the University of Michigan. Currently, he is Professor of Mechanical Engineering and Applied Mechanics at the University of Michigan.

Shigeaki Inada

Shigeaki Inada is associated with Gunma University, Kiryu, Japan. He is Associate Professor of Mechanical Engineering on leave with the University of Michigan as a visiting scholar.

PART **I.4**
Mixing, dispersion, and diffusion

Fig. I.4. Jet diffusion flame transition to turbulence. Note the toroidal vortices outside the laminar flame surface and the initial instability wave in Fig. 1. The inner instability develops into coherent vortices as shown in Fig. 2. The inner vortices then lose their coherence through coalescence and small-scale unorganized vortices form in Fig. 3. (Courtesy of W. Roquemore, L.-D. Chen, J. Seaba, P. Tschen, L. Goss, and D. Trump, *Phys. Fluids* 30, 9 (1987): 2600.)

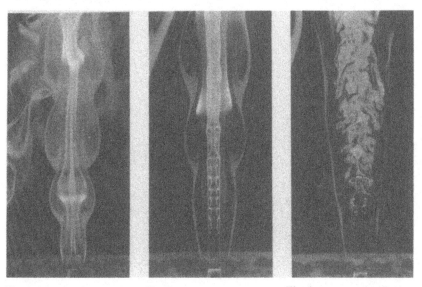

Fig. 1   Fig. 2   Fig. 3

EXPERIMENT 21

Determination of the binary diffusion coefficient in a liquid mixture*

Contributed by

WILLIAM A. WAKEHAM

Principle

The velocity profile in the laminar flow of a fluid through a cylindrical tube is employed to disperse an injected solute. The radial diffusion of the solute in the tube, arising from the radial concentration gradients so created, opposes this dispersion. The combined effect is to produce a solute distribution in the longitudinal direction within the tube that is Gaussian and whose second central moment is related to the mutual diffusion coefficient of the fluid system.

Object

The object of the experiment is the measurement of the mutual diffusion coefficient of a binary liquid mixture. The experiment illustrates how an understanding of fluid mechanics and transport processes can be employed to develop a powerful and simple measurement technique. The aim of the present simple experiment to be described is to determine the diffusion coefficient of the system n-hexane/n-heptane for almost pure n-heptane at 25 °C.

Background

The process of diffusion is the name applied to the relative motion of molecular species in a fluid mixture under the influence of a gradient of concentration, or more completely, of chemical potential. The simplest possible realization of the process is illustrated in Fig. 21.1, where at time $t = 0$ a concentration difference is established in a binary fluid mixture across an interface at $z = 0$. The concentration difference leads to a flux of molecules of species 1, relative to a frame of reference fixed with respect to the laboratory, given by

* The first record of this type of experiment was given by G. I. Taylor in Ref. 3.

157

$$J_1 = D_{12}\frac{\partial c_1}{\partial z} \tag{21.1}$$

and a concomitant flux of species 2 given by

$$J_2 = D_{12}\frac{\partial c_2}{\partial z} \tag{21.2}$$

when there is no volume of mixing of the two species. These fluxes lead to a mixing of the two species, as illustrated in Figs. 21.1(b) and 21.1(c), in such a way as to reduce the concentration gradients. The time evolution of the system is expressed by the second-order differential equation

$$\frac{\partial c_1}{\partial t} = \frac{\partial}{\partial z} D_{12} \left[\frac{\partial c_1}{\partial z} \right] \tag{21.3}$$

which is known as Fick's Law of Diffusion.

The aforementioned process can itself be approximated in the laboratory when the fluids are confined in a vessel and care is taken so that the higher-density fluid mixture is below that of lower density, so that natural convection is avoided. Consequently, the solution of Eq. (21.3), subject to an appropriate set of boundary conditions, allows measurements of the concentration changes of either species as a function of time to be used to determine D_{12}. In practice, this technique of measurement is rather difficult and not a little tedious. Both difficulties arise principally from the fact that the process of diffusion in liquids is very slow. Consequently, it requires several days for the composition of the mixture to change sufficiently to permit its precise measurement, and throughout that time the pressure and temperature in the cell must be maintained constant and uniform. Furthermore, it is always necessary to work with small concentration differences between the two

Fig. 21.1. The process of diffusion.

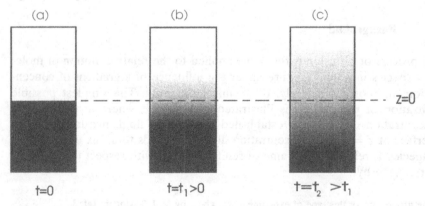

fluid mixtures because otherwise one must account for the concentration dependence of D_{12} in solving Eq. (21.3) and allow for the fact that in any real fluid system there is a volume of mixing. This fact exacerbates the slowness of the entire measurement.

The present experiment provides a means of measuring a diffusion coefficient in a liquid mixture in a period of the order of an hour that is as accurate as most other techniques but which differs totally from them in concept. The process is illustrated schematically in Fig. 21.2 where fluid 1 is shown in laminar flow through a cylindrical tube. The axial velocity component, u_z, in the flow is, of course, parabolic in its dependence on the radial coordinate r, being zero at the tube walls and a maximum on the centerline. If a slug of fluid 2 of infinitesimal axial extent is injected into the tube at a position $z = 0$ at time $t = 0$, then three coupled processes take place. First, the axial velocity transports species 2 downstream at a rate determined by the radial position of fluid elements. This gives rise to two concentration gradients, one in the axial direction and the other in the radial direction, so diffusion of species 2 takes place in these two directions carrying molecules of species 2 from the radial position of their original injection to positions where the axial convective velocity is different. As a consequence, downstream from the point of injection at $z = L$ the concentration distribution of species 2 is shown schematically in Fig. 21.2. That is, the material is not simply distributed along the solid line representing the velocity profile, but has a spread on either side of it.

The extent of the spread, for a given distance L, depends upon the velocity of the flow and the dimensions of the tube but also upon the diffusion coefficient of species 1 and 2, namely, D_{12}. Consequently, if one can determine the spread of species 2 then it is possible to determine D_{12}.

Fig. 21.2. A schematic diagram of the diffusion process in laminar flow.

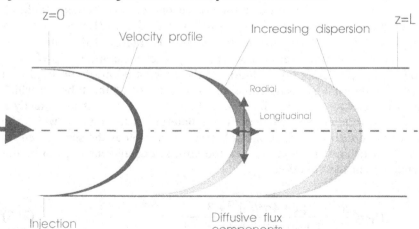

Theory

A simple mass balance for an element of fluid in Fig. 21.2 leads to the following differential equation for the concentration of species 2 at any point in the system

$$
\frac{1}{D_{12}} \frac{\partial c_2}{\partial t} = \nabla^2 c_2 - \frac{2\bar{u}}{D_{12}} \left[1 - \left[\frac{r}{R} \right]^2 \right] \frac{\partial c_2}{\partial z}
\tag{21.4}
$$

in which ∇^2 is the Laplacian operator of a cylindrical polar coordinate system. This equation is solved for an initial injection of solute 2 concentrated in a delta function pulse at $z = 0$ at time $t = 0$ and for the case when there is no penetration of the tube wall by material. One finds that after some time the instantaneous distribution of the concentration of solute 2, averaged over a cross section of the tube, conforms to a Gaussian or normal distribution given by

$$
\bar{c}_2 = \frac{N_2}{\sqrt{2}\,\pi^{3/2}\,R^2\,\mu_2'^{1/2}} \exp \left\{ \frac{-(z - \bar{u}t)^2}{2\mu_2'} \right\}
\tag{21.5}
$$

where N_2 is the number of moles of species 2 injected. The symbol μ_2' denotes the second central moment of the distribution in space and it is related to the diffusion coefficient by the expression

$$
\mu_2' = 2 \left[D_{12} + \frac{\bar{u}^2 R^2}{48\,D_{12}} \right] t
\tag{21.6}
$$

If therefore follows that if $\bar{c}_2$ could be determined within the tube at any instant of time as a function of z then its second central moment could be measured and D_{12} evaluated from it with the aid of Eq. (21.6) and a knowledge of the mean velocity of the flow $\bar{u}$ and the radius of the tube.

The previous arrangement is not especially practicable and it is preferable to determine the temporal evolution of the concentration of species 2 at a specific location $z = L$, usually chosen to be the exit of the tube. In which case, one observes $\bar{c}_2(L,t)$ which, according to Eq. (21.5), is not exactly a normal distribution. Nevertheless, if one determines the time at which the peak of the effluent distribution is observed $\bar{t}$, as well as the second central moment of this distribution σ^2, then the diffusion coefficient D_{12} can be determined from the equation

$$
D_{12} = \frac{R^2}{24\bar{t}} \frac{[1 + 4\sigma^2/\bar{t}^2]^{1/2} + 3}{[1 + 4\sigma^2/\bar{t}^2]^{1/2} + 2\sigma^2/\bar{t}^2 - 1}
\tag{21.7}
$$

in which the radius of the tube is the only additional quantity that needs to be measured.

In the most precise work it is necessary to consider a number of refinements to the theory. However, these same refinements allow an instrument to be designed in such a way that Eq. (21.7) is adequate for the evaluation of the diffusion coefficient. The characteristics of the apparatus described in the next section have been chosen to ensure that the simplest form of the working equation is adequate for many purposes.

Apparatus

A stainless-steel diffusion tube, length approximately 13 m, i.d. 0.8 mm
Liquid chromatograph injection port
Liquid syringe, volume 1 μL
Liquid chromatograph refractive-index detector
Constant-temperature bath
Thermometer
Chart recorder
Glass vessel for liquid reservoir
Nylon connecting tube

Figure 21.3 shows the apparatus assembly.

Procedure

The diffusion tube is wound into a coil of diameter no less than 130 mm on a metal former designed to act as a heat sink. The length of the diffusion tube should be measured and the cross-sectional area πR^2 determined by weighing a section of tube, first empty and then full, of mercury. At one end the diffusion tube is fitted to the chromatograph injection port and at the other to one side of the differential refractive-index detector. In both cases, the junctions should be made with low-dead-volume couplings. The upstream side of the injection port is connected, by means of a flexible tube, to a glass reservoir containing normal heptane. This reservoir should be arranged so that its height can be adjusted to up to 2 m above the level of the diffusion tube since it serves as the gravity feed for the liquid flow. A bypass connection from the reservoir to the reference side of the differential refractive-index detector should be made through a constricted tube to secure a flow rate approximately equal to that through the diffusion tube. The refractive-index-detector output should be connected to a chart recorder.

To begin the experiment, pure heptane is allowed to flow through the diffusion tube, maintained at a constant temperature, as well as the reference system until a stable baseline is obtained on the chart record. The volumetric flow rate should be adjusted using the variable gravity head until it is approximately 1 cm³/min. A sample of 10 percent by volume of n-octane in n-heptane is then manufactured and 1 μL drawn into a syringe. This sample is then injected into the sample port with the syringe and a mark placed on the chart. The chart should be left running continuously and after a period of approximately one hour a peak will be produced on the chart record, similar to that shown in Fig. 21.4, as the dispersed sample of n-octane passes the refractive-index detector. The chart may be stopped after the peak has passed and subjected to analysis.

It is worthy of note that some adjustment of the output level of the refractive-index detector and of the input level of the chart recorder, as well as the chart speed, will be necessary to attain a peak of suitable dimensions for analysis. If the equipment is available the chart recorder may be replaced by computerized data acquisition to permit a more careful analysis.

Analysis and results

The chart record acquired in the experiment is now used to determine the time at which the peak of the distribution occurs $\bar{t}$ and the second central

Fig. 21.3. The apparatus assembly.

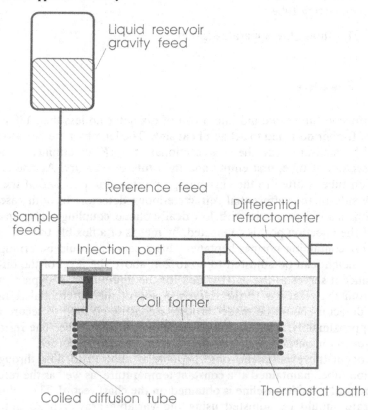

moment σ^2. To determine the former, one simply determines the distance on the chart record from the point of injection to the point of the maximum of the peak. To determine the latter, there are two possible constructions:

1. Determine the height of the maximum of the peak above the baseline and then measure the width of the peak at one half of this height and convert this to an equivalent time $t_{1/2}$. This width is related to σ^2 for a normal distribution by the relationship

$$\sigma^2 = t_{1/2}^2/8ln^2 \tag{21.8}$$

2. Draw tangents to the peak at the point of inflection on each side of it and determine the distance between the points at which the tangents cross the baseline and convert the distance to a time t_b. This time is related to σ^2 for a normal distribution by the relationship

$$\sigma^2 = t_b^2/16 \tag{21.9}$$

The diffusion coefficient can then be evaluated from Eq. (21.7) and a value of approximately

$$D_{12} = 3.4 \times 10^{-9}\, m^2\, s^{-1} \tag{21.10}$$

should be obtained.

Fig. 21.4. A typical diffusion peak with measurement constructions.

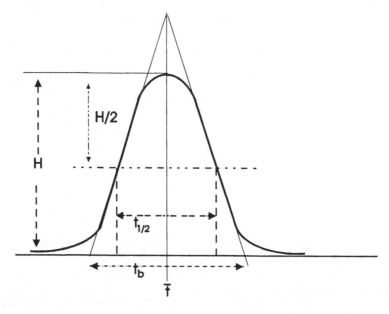

References

1. Alizadeh, A.; Nieto de Castro, C. A., and Wakeham, W. A., *Int. J. Thermophys.* 1 (1980): 243.
2. Alizadeh, A., and Wakeham, W. A., *Int. J. Thermophys.* 3 (1982): 307.
3. Taylor, G. I., *Proc. Roy. Soc.* A219 (1953): 186.

William A. Wakeham

William Wakeham received his Ph.D. degree in physics from the University of Exeter, U.K. He is currently Professor of Chemical Physics and head of the Department of Chemical Engineering and Chemical Technology of Imperial College at the University of London. He has written approximately 180 scientific papers on the thermophysical properties of fluids and intermolecular forces with a special emphasis on the accurate determination of the transport properties over a wide range of thermodynamic states. He is the co-author of two books on intermolecular forces, and of a volume on the measurement, theory, and prediction of the transport properties of fluids.

EXPERIMENT 22

Diffusion distillation: A separation method for azeotropic mixtures

Contributed by
E.-U. SCHLÜNDER

Principle

When distillation occurs in the presence of an inert gas, the separation of the respective components is not only vapor-pressure but also diffusion controlled.

Object

The experiment demonstrates how a binary azeotropic mixture can be separated due to a vapor-diffusion effect, provided that the two components have different diffusivities with respect to the inert gas and, further, the partial pressures of the vapors are kept sufficiently below the total pressure. The diffusion effect disappears when the liquid is evaporated at boiling temperature. This is all well described by the Stefan–Maxwell equation for multicomponent diffusion.

Background

Molecular diffusion in a multicomponent system under isothermal and isobaric conditions is correctly described by the Stefan–Maxwell equations. For steady-state, unidirectional transfer they reduce to

$$\frac{d\tilde{y}_i}{dZ} = \sum_{k=1}^{n} \frac{\tilde{y}_i \dot{r}_k - \tilde{y}_k \dot{r}_i}{\delta_{ik}} \quad \text{with} \quad Z = \frac{\dot{N}z}{An_g}, \quad \dot{r}_i = \frac{\dot{N}_i}{\dot{N}}, \quad \dot{N} = \sum_{i=1}^{n} \dot{N}_i$$

(22.1)

These equations can be simplified for the case of transfer of two species through a third stagnant component ($\dot{N}_3 = 0$). Moreover binary diffusion coefficients of isopropanol–steam and isopropanol–air are nearly the same. Then Eq. (22.1) can be written for components 1, 3:

$$-\delta_{13} \frac{d\tilde{y}_1}{dZ} = \dot{r}_1 - \tilde{y}_1$$

165

$$-\delta_{23}\frac{d\tilde{y}_3}{dZ} = \tilde{y}_3\left(\left(1 - \frac{\delta_{23}}{\delta_{13}}\right)\dot{r}_1 - 1\right) \tag{22.2}$$

Integration past the thickness of the boundary layer S leads to

$$\frac{\tilde{y}^*_{1,c} - \dot{r}_1}{\tilde{y}^*_{1,v} - \dot{r}_1} = \exp(Z^*)$$

$$\frac{\tilde{y}^*_{3,c}}{\tilde{y}^*_{3,v}} = \exp\left[Z^*\left(\frac{\delta_{13}}{\delta_{23}} - \left(\frac{\delta_{13}}{\delta_{23}} - 1\right)\dot{r}_1\right)\right] \tag{22.3}$$

with $Z^* = \dfrac{\dot{N}S}{A n_g \delta_{13}}$

The vapor–liquid equilibria are described by

$$\tilde{y}_i^* = K_i \tilde{x}_i$$

with an equilibrium constant:

$$K_i = \gamma_i \frac{p_i^*(T)}{\rho} \tag{22.4}$$

For calculating the selectivity of the process, the parameter Z^* can be eliminated in Eq. (22.3). With the relation for equilibria Eq. (22.4), the condition

$$\sum_i \tilde{y}_i = 1 \tag{22.5}$$

and the mass balance in the case of total condensation

$$\dot{r}_i = \tilde{x}_{1,c} \tag{22.6}$$

calculation yields the final expression:

$$\left(\frac{1 - K_{2,c} + (K_{2,c} - K_{1,c})\,\tilde{x}_{1,c}}{1 - K_{2,v} + (K_{2,v} - K_{1,v})\,\tilde{x}_{1,v}}\right) = \left(\frac{K_{1,c}\,\tilde{x}_{1,c} - \tilde{x}_{1,c}}{K_{1,v}\,\tilde{x}_{1,v} - \tilde{x}_{1,c}}\right)^{\left(\frac{\delta_{13}}{\delta_{23}} - \left(\frac{\delta_{13}}{\delta_{23}} - 1\right)\tilde{x}_{1,c}\right)} \tag{22.7}$$

Figure 22.1 shows experimental results and calculated profiles of condensate composition. With rising evaporation temperature they tend toward the equilibrium curve. Because of its higher diffusivity in air, water is preferentially transferred in the azeotropic region, which cannot be overcome by normal distillation at boiling temperature. To evaluate the separation process, selectivity and transfer efficiency are essential. Transforming Eq. (22.3) leads to the following relation:

$$Z^* = \ln\left(\frac{\tilde{x}_{1,c}(1 - K_{1,c})}{\tilde{x}_{1,c} - K_{1,v}\,\tilde{x}_{1,v}}\right) \tag{22.8}$$

The effect of temperature on selectivity and mass flux for azeotropic composition ($\tilde{x}_{1,az} = 0.68$) is shown in Fig. 22.2. The curve has a maximum and trend to higher temperatures. Because of convection the measured transfer rates exceed the theoretical values. If condensation at the wall is in evidence, selectivities may arise with higher temperatures. An optimization for industrial applications has to reduce the gap width, which has no influence on condensate composition (according to theory) but some on transfer rate. A desired secondary effect is the arising stability of the system against thermal convection. The limit is given by the demand of strict separation of the evaporating and condensed liquids.

Apparatus

The apparatus consists of a glass beaker (diameter 0.1 m, height 0.2 m), which is heated from below at constant temperature and cooled from above with water vaporizing in an air stream (Fig. 22.3). The liquid mixture is evaporated below the boiling temperature, diffuses through the inert gas layer, is recondensed at the watch glass, and is sampled for analysis. (Losses of the condensed liquid by evaporation in the sampler should be prevented.) The composition of the mixture both on the evaporation and condensation sides are determined with a density meter.

Fig. 22.1. Condensate composition at different evaporation temperatures.

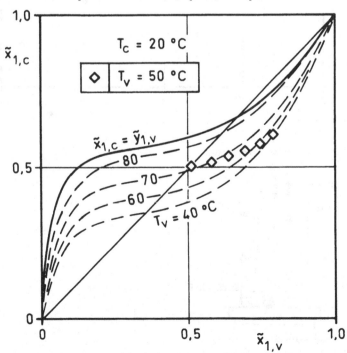

Procedure

The experiments are carried out with isopropanol(1)–water(2) mixtures at mole fractions $\tilde{x}_1$ around the azeotropic composition ($\tilde{x}_1 = 0.68$). There are two experimental procedures: First the temperature will be kept constant at various compositions of the liquid mixture and second the temperature will be varied at constant composition. The volume of sample shall not exceed 5 cm³. The evaporation rate should be measured to compare selectivity and transfer efficiency.

Fig. 22.2. Selectivity and transfer efficiency near the azeotropic composition.

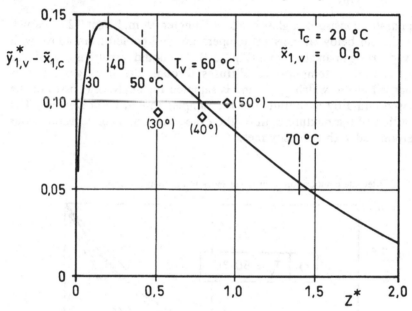

Fig. 22.3. Diffusion distillation apparatus.

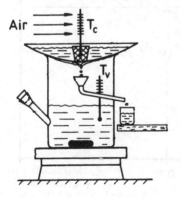

Variation of composition

Start with 150 cm^3 of mixture ($\tilde{x}_1 = 0.5$). The temperature of evaporating liquid is kept constant at $50 \,^\circ\text{C}$. After getting samples from both sides, remove 20 cm^3 of liquid with a syringe and add 30 cm^3 of alcohol, heated before to the desired temperature. Repeat this procedure five times to get a marked change of composition.

Variation of temperature

Add 300 cm^3 of mixture ($\tilde{x}_1 = 0.6$) and get samples of the condensing liquid at different temperatures ($30, 40, 50, 60, 70 \,^\circ\text{C}$), up to the boiling point. During periods of warming, condensed liquid will be sampled separately. Determine the composition of evaporating liquid before and after each run. At higher temperatures experimental results will be influenced by condensation at the wall. Prevent this by a heating tape. At the boiling temperature take samples after a period of time to eliminate diffusion effects.

Suggested headings

(c: condensation side, v: evaporation side)

Time	$T_v(\,^\circ\text{C})$	$T_c(\,^\circ\text{C})$	$\tilde{x}_{1,v}$	$\Delta M_c(g)$	$\tilde{x}_{1,c}$	Z^*	$\tilde{x}_{1,c,\text{calc.}}$

References

1. Fullarton, D.; Schlünder, E.-U., "Diffusion distillation: a new separation process for azeotropic mixtures," *Chem. Eng. Fund.* 2, 1 (1983) 53–65.
2. Fullarton, D.; Schlünder, E.-U., "Diffusion distillation: a new separation process for azeotropic mixtures. (Part I: selectivity and transfer efficiency. Part II: dehydration of isopropanol by diffusion distillation.)," *Chem. Eng. Proc.* 20 (1986) 255–70.

Notation

A_i, B_i, C_i, D_i		constants
A	m^2	interfacial area
K_i		equilibrium constant
$\tilde{M}_i$	kg/kmol	molecular weight
$\dot{N}$	mol/s	molar flux
n_g	mol/m^3	molar density
p_i^*	Pa	vapor pressure of pure species i

p	Pa	pressure
$\dot{r}_i$		relative flux
S	m	distance between liquid interfaces
T	K	absolute temperature
$\tilde{x}_i$		mole fraction in liquid mixture
$\tilde{y}_i$		mole fraction in gas mixture
z	m	distance along the diffusion path
Z		dimensionless distance
δ_{ik}	m²/s	binary diffusivity in the gas
γ_i		activity coefficient

Appendix

(1 = isopropanol, 2 = water, 3 = air)

vapor pressure:
$$\ln(p_i^*/Pa) = A_i + \frac{B_i}{T/°C + C_i}$$

$A_1 = 11.00323,\ B_1 = -2010.33,\ C_1 = 252.636$
$A_2 = 10.19625,\ B_2 = -1730.63,\ C_2 = 233.426$

activity coefficient:
(van Laar equation)
$$\ln\gamma_i = \frac{D_i D_k^2\ \tilde{x}_k^2}{(D_i \tilde{x}_i + D_k \tilde{x}_k)^2}\ (i,k = 1, 2)$$

$D_1 = 2.3405,\ D_2 = 1.1551$

molecular weight: $\tilde{M}_1 = 60$ kg/kmol; $\tilde{M}_2 = 18$ kg/kmol

liquid density: $\rho(20\ °C)/(kg/m^3) = 1030.7 - 244.3\ x_1$
$0.4 < x_1 < 1,\ x_1$ = mass fraction

diffusion coefficient: $\delta_{13}/(m^2/s) = 8.59 \cdot 10^{-6}(T/T_o)^{1.91}$
$\delta_{23}/(m^2/s) = 22.81 \cdot 10^{-6}(T/T_o)^{1.81}$
$(T_o = 273.15K)$

Ernst-Ulrich Schlünder

Professor Ernst-Ulrich Schlünder is head of the Institut für Thermische Verfahrenstechnik (thermal process engineering) at the University of Karlsruhe, Germany. He graduated in 1958 from the Technische Hochschule Darmstadt, Germany, and was an instructor in 1962 and 1964 at TH Darmstadt and TH Hannover, respectively. He was head of the Heat Transfer Department of the Max Planck Institut für Strömungsforschung (fluid mechanics research) at Göttingen until 1967, and since then has been professor at Karlsruhe.

PART I.5
Radiation

Fig. I.5. Infrared photograph of the walls of a heat exchanger. The picture was taken in total darkness and only infrared is recorded. Heat patterns invisible to the eye can be observed, and by calibrating against thermocouples, approximate temperatures can be determined. For a given exposure, areas of equal grayness will be at equal temperatures. By use of time-lapse photography, the movement of the blue-brittle range across the heat exchanger can be observed during cycling. (Courtesy of H. L. Gibson, *Photography by Infrared*, Wiley, New York, 1939.)

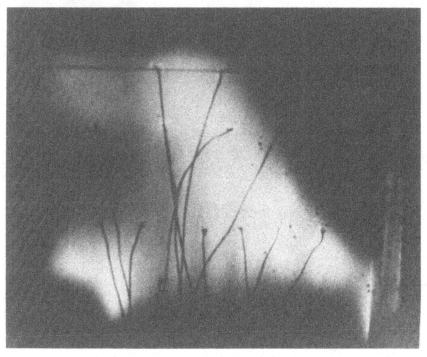

EXPERIMENT 23

Determination of the emissivity of a heated copper surface

Contributed by
ROSS LEONARD JUDD

Principle

The rate at which heat is transferred from the surface of a heated object will become equivalent to the rate at which heat is generated within the object when steady-state conditions have been attained. If the associated conduction and convection heat-transfer effects can be eliminated, radiation heat transfer will become the only means of transferring the heat.

Object

It is an easy matter to model the heat-exchange process that takes place between a heated sphere and a spherical shell in which it is concentrically located because of the simplicity of the geometry. If the heat flowing along the support by which the heated sphere is suspended is negligible and most of the air has been extracted from the space between the heated sphere and the spherical shell, radiation heat transfer becomes the only mechanism capable of transferring the heat generated within the sphere. The purpose of this experiment is to determine the emissivity of a copper surface by substituting the temperatures of the heated sphere and the glass shell corresponding to steady-state conditions into the radiation heat-transfer equation and performing an analysis that yields the value of emissivity.

Background

Figure 23.1 depicts a copper sphere of radius R_{cs} suspended within a spherical glass jar of radius R_{gi}. The sphere is attached to the cover plate by a thin-walled stainless-steel tube which prevents any significant heat transfer by conduction. The air in the space between the heated sphere and the spherical shell is rarefied to such an extent that natural convection is almost completely eliminated. Under these conditions, the rate at which heat is generated electrically in the cartridge heater at the center of the sphere, q_e, can be related

173

174   R. L. Judd

to the rate of radiation heat exchange between the copper sphere, considered to be a small gray object, and the spherical shell, considered to be a large black enclosure, according to the relationship

$$q_e = 4\pi R_{cs}^2 \varepsilon_{cs}\sigma\left[T_{cs}^4 - T_{gj}^4\right] + q_c \tag{23.1}$$

where ε_{cs} is the emissivity of the copper sphere, $\sigma = 0.1718 \times 10^{-8}$ BTU/hr ft^2 °R^4 is the Stefan Boltzman constant, T_{cs} is the temperature of the copper sphere, and T_{gj} is the temperature of the glass jar. The term q_c on the right-hand side of Eq. (23.1), which is a very small fraction of the rate of the heat generation q_e, accounts for the transfer of heat through the rarefied atmosphere in the space between the copper sphere and the glass jar by natural convection at reduced pressure. The theory of Kyte, Madden, and Piret,[1] which is represented by the curves plotted in Fig. 23.2, enables the prediction of the value of q_c as a function of the temperature of the copper sphere T_{cs} and the pressure of the air p.

Under steady-state conditions, all of the energy that is generated within the copper sphere is transferred to the glass jar, which in turn transfers it to the surroundings by radiation and natural convection. If we assume that the wall of the glass jar is so thin that there is no significant temperature variation

Fig. 23.1. Arrangement of the experimental apparatus.

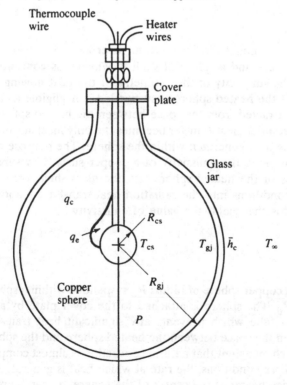

within it, an energy balance applied at the outer surface of the glass jar leads to the equation

$$q_e = 4\pi R_{gj}^2 \sigma [T_{gj}^4 - T_{\infty}^4] + 4\pi R_{gi}^2 \bar{h}_c (T_{gj} - T_{\infty}) \qquad (23.2)$$

where T_{∞} is the temperature of the surroundings and $\bar{h}_c$ is the natural-convection heat-transfer coefficient existing at the outer surface of the glass jar, which can be predicted by means of the empirical correlation

$$\bar{h}_c = 0.27 \left[\frac{(T_{gj} - T_{\infty})}{R_{gj}} \right]^{0.25} \qquad (23.3)$$

so substitution of Eq. (23.3) into Eq. (23.2) leads to the equation

$$q_e = 4\pi R_{gj}^2 \sigma [T_{gj}^4 - T_{\infty}^4] + 4\pi \times 0.27 R_{gj}^{1.75} (T_{gj} - T_{\infty})^{1.25} \qquad (23.4)$$

The only unknown in Eq. (23.4) is the temperature of the glass jar T_{gj}, which can be determined by trial and error. Substitution of the value of T_{gj} obtained from Eq. (23.4) into Eq. (23.1), in which the unknowns are T_{gj} and ε_{cs}, enables the emissivity of the copper surface ε_{cs} to be determined.

Fig. 23.2. Heat-transfer rate versus copper-sphere temperature for a variety of pressures.

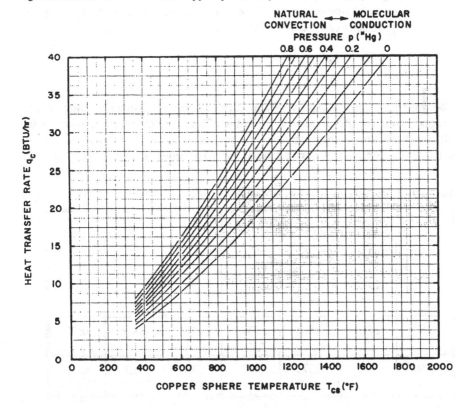

Apparatus

Copper sphere ($R_{cs} = 0.75''$) containing a 150 W cartridge heater
Spherical glass jar ($R_{gj} = 5.65''$) fitted with a cover plate
Variac for controlling the power generated in the cartridge heater q_e
Wattmeter for measuring the power generated in the cartridge heater q_e
Heise gauge for measuring the pressure of the rarefied atmosphere p
Potentiometer for reading the EMF of the thermocouple measuring T_{cs}
Thermometer for measuring the temperature of the surroundings T_∞

Procedure

Reduce the pressure inside the glass jar to less than 1" Hg and set the electrical power generation to the appropriate level.
Monitor the temperature indicated by the thermocouple located in the copper sphere at five-minute intervals until steady state is attained.
Take a complete set of data consisting of the pressure of the rarefied atmosphere in the spherical jar, the electrical power generation, the copper-
sphere temperature, and the temperature of the surroundings.
Determine the emissivity of the copper sphere in accordance with the solution technique outlined before. Four sets of data comprising a typical
experiment can easily be obtained in a single laboratory session.

Results

The values of the emissivity of the copper sphere ε_{cs} corresponding to the
experimental results presented in Tables 23.1 and 23.2 can be seen to be
different although they obviously agree among themselves. The mean
value of the emissivity derived from the experimental results presented in

Table 23.1. *Representative experimental results*

p	q_e	T_{cs}	T_∞	T_{gj}	q_c	ε_{cs}
"Hg	BTU/hr	°R	°R	°R	BTU/hr	
0.50	204.8	1386	537	574	24.8	0.60
0.50	273.1	1508	537	584	28.8	0.58
0.50	307.3	1572	537	589	31.1	0.55

Table 23.1 is 0.577 with a standard deviation of 0.025 whereas the mean value of the emissivity derived from the experimental results presented in Table 23.2 is 0.713 with a standard deviation of 0.017. This anomalous situation occurred because air was readmitted to the glass jar at the conclusion of an experiment while the copper sphere was still warm. Whenever this happened, an oxide coating would develop on the copper surface and/or some of the oxide coating on the surface would flake off. Reference 2 presents a graph that shows that the emissivity of oxidized copper can vary from 0.50 to 0.90 in the temperature range in which the experiments were conducted, depending upon the degree of oxidization. Accordingly, in as much as the surface condition almost always changed from experiment to experiment, it may be concluded that the values obtained are reasonable. As a consequence, it would appear that the model underlying the analysis and the experimental technique employed is valid.

Suggested headings

Constants:

$R_{cs} = $ _____ ; $R_{gj} = $ _____ ; $T_{\infty} = $ _____

IV		MV		q_c	ε_{cs}
p	q_e	T_{cs}	T_{gj}		

Table 23.2. *Representative experimental results*

p	q_e	T_{cs}	T_{∞}	T_{gj}	q_c	ε_{cs}
"Hg	BTU/hr	°R	°R	°R	BTU/hr	
0.23	162.2	1273	540	570	10.8	0.71
0.23	221.9	1386	540	579	12.8	0.69
0.23	307.3	1489	540	592	15.2	0.72
0.23	366.7	1552	540	601	17.0	0.73

References

1. Kyte, J. R.; Madden, A. J., and Piret, E. L., "Natural convection heat transfer at reduced pressure," *Chemical Engineering Progress* 49, 12 (1953): 653–62.
2. White, F. M., *Heat Transfer*, Addison-Wesley, Reading, MA, 1984.

Ross L. Judd

Professor Judd is on the teaching staff at McMaster University. Professor Judd received his Ph.D. in mechanical engineering from the University Michigan in 1968. From 1958 to 1961, he was employed by the Civilian Atomic Power Department of Canadian General Electric. His current research interests are the investigation of the bubble-nucleation phenomenon and the influence of surface microstructure on the nucleate-boiling heat-transfer process, the effect of two-phase flow on the stability of tube bundles, and the role of heat transfer in metal cutting and machining.

PART I.6
Heat pipes and exchangers

Fig. I.6. Cross-flow heat exchanger, both fluids unmixed. To ensure that the shell-side fluid will flow across the tubes and thus induce higher heat transfer, baffles are placed in the shell as shown. (Courtesy of D. Butterworth, *Harwell Heat Transfer and Fluid Flow Service*, published by The Slide Centre Ltd.)

EXPERIMENT 24
Performance characteristics of an annular heat pipe

Contributed by
A. FAGHRI

Principle

The annular heat-pipe design can significantly increase the heat capacity per unit length compared to conventional cylindrical heat pipes due to the capillary forces generated in the wick material on the inner pipe.

Object

The primary objective of this experiment is to compare the maximum heat transport (capillary) limit of the annular heat pipe to that of a conventional cylindrical heat pipe with the same outer diameter and wick structure. A secondary objective is to examine the temperature distributions on the inside and the outside of the annular heat pipe. Finally, the problem of condensate leakage between the inner wick and the outer wick will be addressed.

Apparatus

The annular heat pipe, as shown in Fig. 24.1, consists of two concentric pipes of unequal diameters attached by means of end caps, which create an annular vapor space between the two pipes. Wick structures are placed on both the inner surface of the outer pipe and the outer surface of the inner pipe (Fig. 24.2). Axial grooves were chosen in this experiment since no special procedures are needed for installation, but any type of wick can be used. The space inside the inner pipe is open to the surroundings. An increase in performance is expected as a result of the increase in surface area exposed for the transfer of heat into and out of the pipe, and the increase in the cross-sectional area of the wick inside the pipe. For a valid comparison between the annular and conventional heat pipes, the length, external diameter, pipe thickness and material, working fluid, and wick structure must be the same. Design summaries of the annular heat pipe, the conventional heat pipe, and the heater

181

Fig. 24.1. Concentric annular heat pipe.

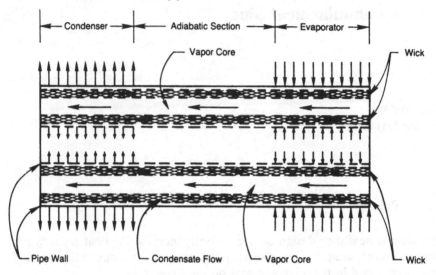

Fig. 24.2. Concentric annular heat-pipe design concept.

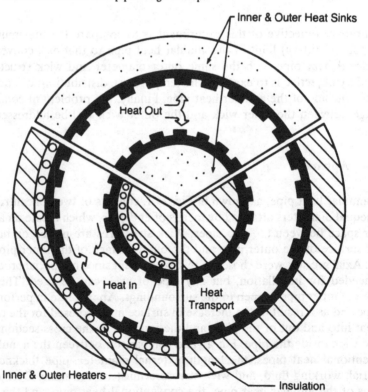

and heat-sink assemblies are given in Tables 24.1–24.4. The schematic of the experimental setup is presented in Fig. 24.3.

Heat is input to the annular heat pipe by two heaters. The inner heater assembly consists of a heater rod coiled around a core of insulation, which is slid into the inner pipe. The outer heaters for the annular heat pipe and the conventional heat pipe are identical heater rods wrapped spirally around the outer pipes and cemented in place with electric-heater cement. This prevents the heater from expanding away from the pipe as it is heated. Heat is extracted from the annular heat pipe with an inner and an outer heat sink. The inner heat sink is a pipe with a solid end cap on one end, an end cap with two holes on the other end, and a solid internal baffle with a large hole to ensure that the coolant travels the entire length of the heat sink. The outer heat sink for the annular and the conventional heat pipes is 1/4-in.-outer-diameter soft copper tubing tightly wrapped around the outer pipe' Thermocouples are placed to measure the inlet and outlet coolant temperatures, and flow meters measure the flow rate of the coolant, so the heat extracted from each heat sink can be calculated as $Q = \dot{m} C_p (T_{out} - T_{in})$. A centrifugal pump fed by a constant-head water tank provides a constant flow rate to the heat sinks.

Thermocouples are also placed on the inner and outer pipes' surfaces in

Table 24.1. *Design summary of the concentric annular heat pipe*

Materials			
Outer pipe	Copper		
Inner pipe	Copper		
End caps	Copper		
Working Fluid	Water		

Dimensions			
		Outer pipe	Inner pipe
Total Length	973 mm		
Evaporator Length	300 mm	OD 50 mm	29.7 mm
Adiabatic Length	473 mm	ID 46.6 mm	25.4 mm
Condenser Length	200 mm	Tw 1.7 mm	2.15 mm

Grooves		
	Outer pipe	Inner pipe
Number	120	97
Width	0.5 mm	0.5 mm
Depth	0.5 mm	0.5 mm
Total groove volume 54.25 cc		

Fluid Inventory	
Quality	Distilled water
Quantity	68 cc filled at 21°C

184   A. Faghri

Table 24.2. *Design summary of conventional pipe*

Materials

Outer pipe	Copper
End caps	Copper
Working Fluid	Water

Dimensions

Total Length	973 mm		
Evaporator Length	300 mm	OD	50 mm
Adiabatic Length	473 mm	ID	46.6 mm
Condenser Length	200 mm	Tw	1.7 mm

Grooves

Number	120
Width	0.5 mm
Depth	0.5 mm
Total groove volume	30 cc

Fluid Inventory

Quality	Distilled water
Quantity	41 cc filled at 21°C

Table 24.3. *Heater assembly design specifications*

Dimensions	Inner Assembly	Outer Assembly
Length	300 mm	300 mm
OD	25.4 mm	...
ID	...	50.4 mm
Surface Area	15959.3 mm^2	32142.3 mm^2
Heating Element		
Length	2.24 m	2.38 m
Diameter	2.36 mm	4.77 mm
Voltage	220 v	220 v
Max. Power	1200 w	4800 w
Assembly Base		
Material	Aremcolox 502-600	
Length	300 mm	
OD	19.05 mm	
Thermocouples		
NBS Type	T (30 gauge)	T (28 gauge)
# of thermocouples	7	7

the locations shown in Fig. 24.3 to determine when the capillary limit is reached. The capillary limit occurs when the wick can no longer supply liquid to the evaporator section at a rate equal to the rate of evaporation. The result is that the liquid in the wick does not completely wet the evaporator section due to premature evaporation, which is called evaporator dryout. The temperature of the evaporator then increases dramatically until it is no longer safe to operate the heat pipe. For safety, the capillary limit is defined as when the temperature difference between the center adiabatic section and the evaporator end cap exceeds 12 °C.

Procedure

Begin operation of the heat pipe being tested by slowly increasing the heat input to the heaters. Allow the heat-pipe temperatures to reach steady state before increasing the heat input. The operating temperature (center adiabatic) should be maintained at approximately 50 °C at all times for repeatability. The operating temperature is controlled by the flow rate and inlet temperature of the coolant flowing in the heat sinks. If the flow rate is increased or the inlet temperature is decreased, the operating temperature of the heat pipe is decreased, and vice versa. Record the temperature distribution along the heat pipe, the heat input to each heater, and the heat output to each heat

Table 24.4. *Heat-sink design specifications*

Dimensions	Inner Sink	Outer Sink
Length	219 mm	200 mm
OD	25.4 mm	63.5 mm
ID	. . .	50.4 mm
Surface Area	17475 mm^2	31667 mm^2
Material		
Outer pipe	Copper (tw = 1.27 mm)	1/4 in OD Copper tubing
Baffle	Copper	. . .
Working fluid	Water	Water
Thermocouples		
NBS Type	T (30 gauge)	
# of thermocouples	8	
Flow Meter		
Max. flow rate	906 ml/min.	1812 ml/min.
Min. flow rate	24 ml/min.	48 ml/min.
Fluid	Water	Water

Fig. 24.3. Experimental setup.

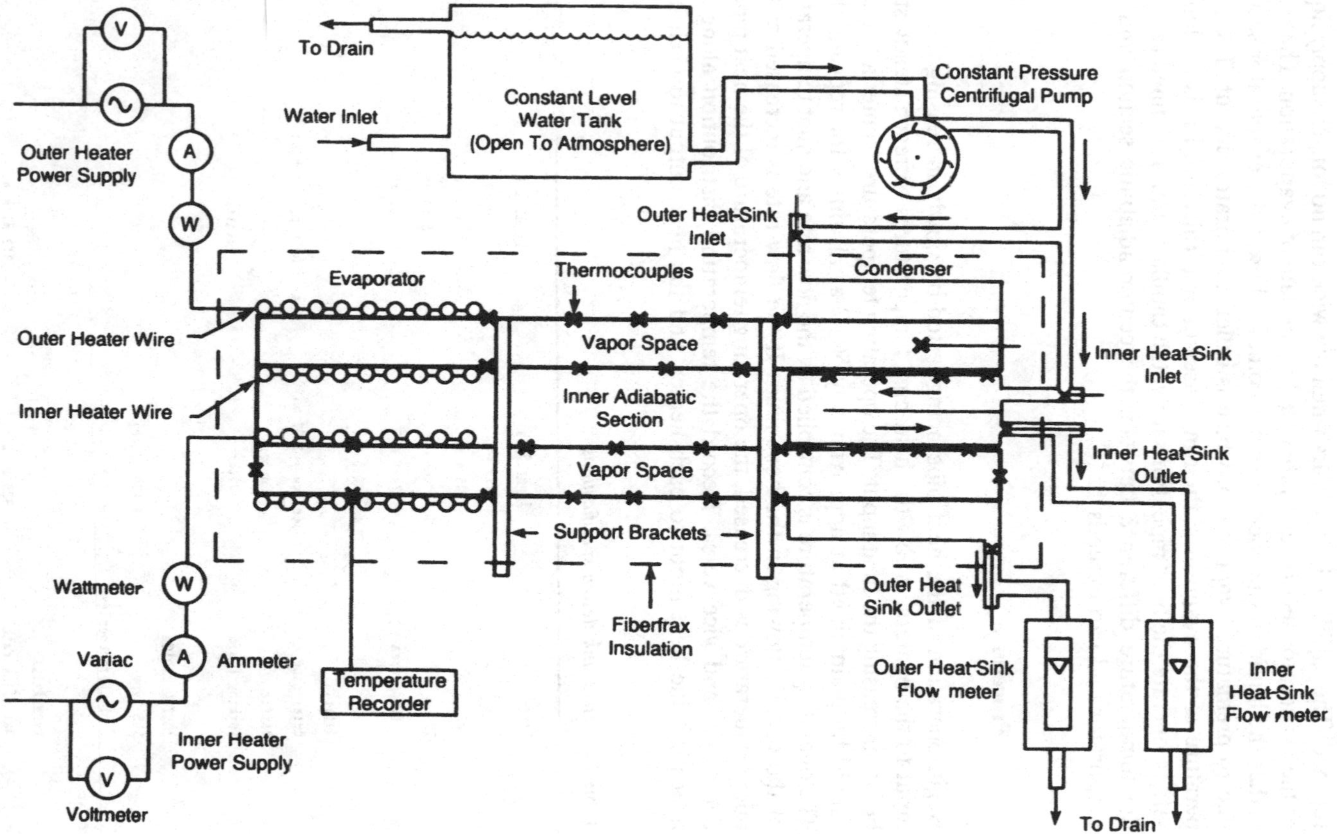

sink. Note and compare the heat input at which each heat pipe reaches the capillary limit. For the annular heat pipe, compare the heat input to and extracted from the inner wall. Compare the same for the outer wall. If the heat input to and heat extracted from the inner pipe are not equal, a meniscus has been formed in the condenser section where the inner pipe and the end cap are joined, which allows part of the working fluid that condenses onto the inner pipe to drain down to the outer pipe.

Capillary-limit calculation

In an annular heat pipe operating under steady conditions, the sum of the pressure changes in the closed-cycle system may be described by the following relation:

$$2[P_v(z_{ref}) - P_v(z)] + [P_v(z) - P_{L,I}(z)] + [P_v(z) - P_{L,O}(z)] +$$

$$[P_{L,I}(z) - P_{L,I}(z_{ref})] + [P_{L,I}(z_{ref}) - P_v(z_{ref})] + [P_{L,O}(z_{ref}) - P_v(z_{ref})] +$$

$$[P_{L,O}(z) - P_{L,O}(z_{ref})] = 0 \tag{24.1}$$

The capillary pressure P_C is defined as the pressure at the vapor side of the liquid interface minus that at the liquid side. It is assumed that the z_{ref} is located at z_{min} where the capillary pressure is minimum and equal to zero. This reduces Eq. (24.1) to

$$P_{C,I}(z) + P_{C,O}(z) = 2\Delta P_v(z - z_{min}) + \Delta P_{L,I}(z_{min} - z) + \Delta P_{L,O}(z_{min} - z) \tag{24.2}$$

For the annular heat pipe, the maximum capillary pressure at the inner and outer walls are

$$P_{C,max,I} = \frac{2\sigma}{r_{c,I}} \tag{24.3}$$

$$P_{C,max,O} = \frac{2\sigma}{r_{c,O}} \tag{24.4}$$

The force balances for the liquid flow in the inner and outer wall grooves are

$$\frac{dP_{L,I}}{dz} = -\frac{4\tau_{L,I}}{D_{h,L,I}} \pm \rho_L g \sin\theta \tag{24.5}$$

$$\frac{dP_{L,O}}{dz} = -\frac{4\tau_{L,O}}{D_{h,L,O}} \pm \rho_L g \sin\theta \tag{24.6}$$

Equations (24.5) and (24.6) can be represented in terms of the local axial heat fluxes Q_I and Q_O for the inner and outer walls, respectively.

$$\frac{dP_{L,I}}{dz} = -F_{L,I}Q_I \pm \rho_L g \sin\theta \tag{24.7}$$

$$\frac{dP_{L,O}}{dz} = -F_{L,O}Q_O \pm \rho_L g \sin\theta \tag{24.8}$$

The functions F are defined in the following way:

$$F_{L,I} = \frac{v_L}{K_I A_{w,I} h_{fg}} \tag{24.9}$$

$$F_{L,O} = \frac{v_L}{K_O A_{w,O} h_{fg}} \tag{24.10}$$

Applying the conservation of axial momentum to the vapor flow between the concentric pipes, one obtains the following relationship provided that the flow is laminar:

$$A_v \frac{dP_v}{dz} = -\tau_{v,I}(\pi D_I) - \tau_{v,O}(\pi D_O) - A_v \frac{d\rho_v w_v^2}{dz} + (\dot{m}_I + \dot{m}_O)w_v \tag{24.11}$$

Since the mass flux of the vapor is related to the axial heat flux at the same $z (Q = \rho_v w_v A_v h_{fg})$, Eq. (24.11) can be presented in the following form when the last term on the right-hand side is neglected:

$$\frac{dP_v}{dz} = -F_{v,av}Q - E_v \frac{dQ^2}{dz} \tag{24.12}$$

where

$$F_{v,av} = \frac{2(f_{v,av}Re_v)v_v}{A_v D_{h,v}^2 h_{fg}} \tag{24.13}$$

$$f_{v,av} = \frac{\tau_{v,av}}{\rho_v w_v^2/2} = \frac{D_O \tau_{v,O} + D_I \tau_{v,I}}{(D_O + D_I)\rho_v w_v^2/2} \tag{24.14}$$

$$Re_v = \frac{\rho_v w_v D_{h,v}}{\mu_v} \tag{24.15}$$

$$D_{h,v} = D_O - D_I \tag{24.16}$$

$$E_v = \frac{1}{A_v^2 \rho_v h_{fg}^2} \tag{24.17}$$

Substituting $\Delta P_L(z - z_{min})$ and $\Delta P_v(z - z_{min})$ from Eqs. (24.7), (24.8), and (24.12) into Eq. (24.2) and neglecting the effects of gravity results in the following equation:

$$2\sigma\left(\frac{1}{r_{c,I}} + \frac{1}{r_{c,O}}\right) = 2F_{v,av}\int_0^L Q dz + \int_0^L (F_{L,I}Q_I + F_{L,O}Q_O)dz \tag{24.18}$$

The previous relation simplifies, if one assumes that the geometry and dimensions of the grooves on the inner pipe and the outer pipe are the same, as well as the same heat input to the inner and outer walls, that is, $F_{L,I} = F_{L,O} = F_{L,av}$ and $Q_I = Q_O = Q/2$.

$$\int_0^L Q\, dz = \frac{2\sigma\left(\dfrac{1}{r_{c,I}} + \dfrac{1}{r_{c,O}}\right)}{2F_{v,av} + F_{L,av}} \qquad (24.19)$$

The maximum heat transport for a conventional heat pipe is given by the following equation:

$$\int_0^L Q\, dz = \frac{2\sigma/r_c}{F_v + F_L} \qquad (24.20)$$

A comparison of the capillary limit of the axially grooved annular heat pipe to a conventional heat pipe using the previous analysis shows an increase of 80 percent using water as the working fluid at 100 °C with the dimensions of the pipes and grooves as specified in Tables 24.1 and 24.2.

Suggested headings

Inner pipe						
V	A	Q_{in}	$\dot m$	ΔT(coolant)	Q_{out}	Q_{out}/Q_{in}

Outer pipe						
V	A	Q_{in}	$\dot m$	ΔT(coolant)	Q_{out}	Q_{out}/Q_{in}

Overall	
Q_{total}	ΔT(evap.-adiab.)

190   *A. Faghri*

References

S. W., *Heat Pipe Theory and Practice*, Hemisphere Publishing, New York, 1976.
2. Dunn, P. D., and Reay, D. A., *Heat Pipes*, 3rd ed., Pergamon, New York, 1982.
3. Faghri, A., "Performance characteristics of a concentric annular heat pipe: Part II – vapor flow analysis," *ASME J. Heat Transfer* 111 (1989): 851–7.
4. Faghri, A., and Thomas, S., "Performance characteristics of a concentric annular heat pipe: Part I – experimental prediction and analysis of the capillary limit," *ASME J. Heat Transfer* 111 (1989): 844–50.

Notation

A	cross-sectional area
C_p	specific heat at constant pressure
D	diameter
f	coefficient of friction
g	gravitational acceleration
h_{fg}	heat of vaporization
K	wick permeability
L	heat-pipe length
$\dot{m}$	mass flow rate
P	pressure
ΔP	pressure difference
Q	axial heat flux
r_c	effective capillary radius
Re	Reynolds number
t	wall thickness
ΔT	temperature difference
w	velocity
W	groove width
z	axial distance along the heat pipe

Subscripts

av	average
C	capillary
h	hydraulic
I	inner wall
L	liquid phase
max	maximum
min	minimum
O	outer wall
ref	reference
v	vapor phase
W	wick

Greek symbols

δ	groove depth
ε	porosity
θ	inclination angle
μ	viscosity
ν	kinematic viscosity
ρ	density

σ   surface tension
τ   shear stress

Amir Faghri

Dr. Faghri joined the Wright State University faculty in 1982 following a visiting professorship at the University of California at Berkeley where he taught heat transfer and energy courses in the Department of Mechanical Engineering. He was responsible for the development of the thermal-sciences laboratories at Wright State University. He is well known for his expertise in the numerical and experimental analysis of falling liquid films and of heat pipes. He has more than one hundred archival publications and two patents to his credit.

EXPERIMENT 25
Heat transfer during drop formation and release

Contributed by
J. MADDREN *and* E. MARSCHALL

Principle

The resistance to heat transfer in a liquid–liquid direct-contact heat exchanger is found to exist mostly in the dispersed phase. An exception is the drop formation and release process, where the resistances to heat transfer in the continuous phase and the dispersed phase are often of the same order of magnitude. Temperature measurements combined with high-speed motion photography allow one to determine the dispersed-phase and continuous-phase resistances as well as the heat-transfer efficiency.

Object

The objective of this experiment is to determine the heat-transfer efficiency of the drop formation process, and to find the internal and external heat-transfer coefficients immediately following release. The liquid–liquid system described here uses hot water as the continuous phase and cold AMSCO petroleum solvent (oil) as the dispersed phase.

Background

Most liquid–liquid direct-contact heat exchangers consist of a vertical spray column. The dispersed phase, which has a lower density than the continuous phase, is injected at the bottom of the column through a nozzle or a set of nozzles and flows upward in the column in the form of drops. The continuous phase is injected at the top of the column and flows downward.

The problem of finding a nonintrusive method to measure the internal temperatures of a developing drop has not been solved. Some researchers have attempted to measure the average drop temperature in the free-rise section of the spray column by effectively changing the column height and measuring the temperature of the dispersed-phase fluid after coalescence. The heat transfer during formation was then found by extrapolating this data

192

to a column of zero height. The technique outlined here uses a single micro-thermocouple that is positioned parallel to the flow direction and pierces the drop as it rises to obtain dispersed-phase temperatures. The stationary thermo-couple records "temperature traces," which depend on the axial and radial position of the thermocouple junction. Examples of temperature traces ob-tained from a drop rising from the dispersed-phase nozzle are shown in Fig. 25.1. In conjunction with the interface motion, which is obtained from a high-speed movie, the temperature-trace data can be integrated to obtain drop temperatures as a function of position and time. The temperature and flow fields for each drop are assumed to be identical for the duration of the ex-periment, depending only on the height of the drop above the nozzle. Also, isotherms are assumed symmetric with respect to the vertical axis of the drop.

The dispersed phase is injected into the column through a single nozzle. The angle of incidence between the thermocouple probe axis and the direc-tion normal to the liquid–liquid interface must be sufficiently small so that the drop motion is not affected. Therefore data at large radial dimensions cannot be obtained. The probe must be positioned above the rest drop and below the height at which the drop deviates from a one-dimensional path due to wake shedding (about one to two drop diameters of free translation). These limits define the region within which temperature data is obtained at

Fig. 25.1. Temperature traces. (*Note*: the time scale is not the same for each trace.)

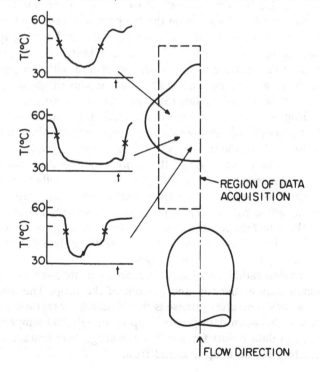

a spacing of approximately 0.25 mm in both the radial and axial directions. That region is indicated in Fig. 25.1.

The temperature-trace data are smoothed with a cubic spline algorithm to obtain continuous first and second derivatives of the temperature as a function of time. The location of the interface is determined by assuming the second derivative is zero and checking for discontinuities in the first derivative due to the differences in physical properties of the dispersed and continuous phases. That is, at the interface the following conditions are assumed to hold:

$$\tau_c \left(\frac{dT(t)}{dt} \right)_c = \tau_d \left(\frac{dT(t)}{dt} \right)_d \qquad (25.1)$$

$$k_c \frac{dT_c}{dz} = k_d \frac{dT_d}{dz} \qquad (25.2)$$

where $T(t)$ is the thermocouple temperature, t is the time, τ_c and τ_d are the thermocouple time constants in the continuous phase and the dispersed phase, respectively, k_c and k_d are the thermal conductivities of the continuous phase and the dispersed phase, respectively, and T_c is the continuous-phase temperature and T_d is the dispersed-phase temperature at the interface.

Even though the discontinuities in the first temperature derivatives do not appear directly in the data due to the finite size of the thermocouple junction and the use of the curve-fitting algorithm, they can be found within the interval of time it takes for the probe junction to pass from one fluid to the other. Therefore, the interface location on the temperature-trace data can be identified and the data corrected for the thermal lag of the probe using appropriate correlations for the calculation of the time constants.

In order to be able to correlate temperatures measured with the position and shape of the drops at a given time, high-speed movies of the drop formation and release process are made. (Still photos of the formation and release of an oil drop in water are shown in Fig. 25.2. Lines inside the drop are the result of a photochromic flow-visualization experiment. They indicate the nature of the flow field inside the drop.) The frame rate of the high-speed movie must be sufficient to interpolate the motion of the drop interface between frames. (A film speed of 400 frames per second is usually sufficient.) For each frame discrete points along the interface are digitized and a nonsmoothing cubic spline routine with appropriate boundary conditions is used to represent the interface continuously. Interface velocities with respect to the probe can then be calculated and temperature-trace data can be correlated with time in the drop formation process.

Temperatures at large radial dimensions are obtained through interpolation of the collected data at the top and bottom of the drop. The average drop temperature at any time after release is then found by integration using linear shape functions between data points. Approximately 400 temperature traces provide enough data points for a reliable average-temperature calculation. The heat rate q to the drop is found from

Fig. 25.2. Flow field inside drop during formation and release.

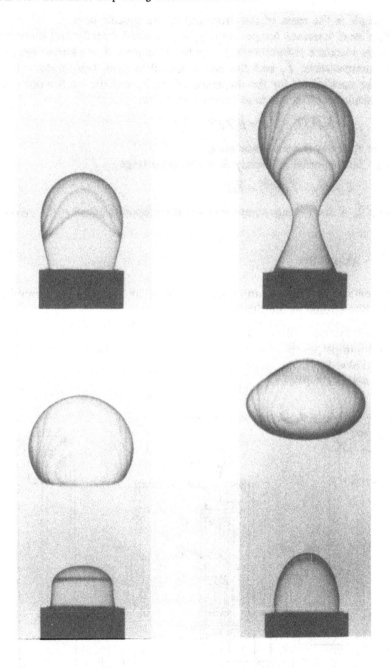

$$q = \rho V c_p d\overline{T}_d/dt \qquad (25.3)$$

where ρV is the mass of the drop and c_p the specific heat.

The local interface temperature T_i is measured directly and therefore the average interface temperature $\overline{T}_i$ can be evaluated. If one knows the average drop temperature $\overline{T}_d$ and the continuous-flow bulk temperature $\overline{T}_c$, heat-transfer coefficients for the dispersed phase h_d and the continuous phase h_c are calculated from the heat balance equation

$$q = h_d A_s(\overline{T}_i - \overline{T}_d) = h_c A_s(\overline{T}_c - \overline{T}_i) \qquad (25.4)$$

where A_s is the drop surface area.

The heat-transfer efficiency E is obtained from

$$E = (\overline{T}_d - \overline{T}_{di})/(\overline{T}_c - \overline{T}_{di}) \qquad (25.5)$$

where $\overline{T}_{di}$ is the average temperature of the dispersed phase at the nozzle exit.

Apparatus

A schematic of the test apparatus is shown in Fig. 25.3. The numbers in parentheses refer to the figure.

Heating (1) and cooling (7) apparatus
Liquid pumps (2, 8)
Head tanks (3, 9)
Flow meters (4, 10)

Fig. 25.3. Schematic of experimental apparatus.

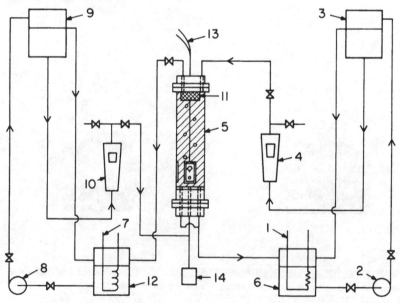

Heating and cooling reservoirs (6, 12)
Spray column, 63.5-mm i.d. (5)
Stainless-steel nozzle, 4.57-mm i.d. and 6.35-mm o.d.
Coalescence chamber (11)
Microthermocouple and positioning device (13)
Pressure transducer (14)

In addition, the following equipment is needed:

A/D data acquisition system
High-speed camera
Digitizing system
Computer

Procedure

Assemble the apparatus as in Fig. 25.3. The nozzle should be insulated from
the continuous phase before injection to the spray column.
The thermocouple probe is constructed from fine gauge wire 0.013 mm in
diameter and is supported by a mechanism which can move with a
high degree of accuracy in both the horizontal and vertical directions
(see Fig. 25.4). The thermocouple should be referenced to the oil
flow in the nozzle.
Establish steady-state conditions within the spray column. Calibrate the probe
position. Position the probe in the path of the dispersed-phase drop
and record the drop temperatures. Repeat for every location within
the region of data acquisition.
After all temperature data is recorded and stored, take a high-speed movie
of the drop formation and release process.

Figure 25.5 shows temperature fields in a developing drop: during formation (Fig. 5a), at release (Fig. 5b), and after release (Figs. 5c and 5d).

Suggested headings

Constants:

$\rho =$ _____; $V =$ _____; $c_p =$ _____

$\overline{T}_{di} =$ _____; $\overline{T}_c =$ _____

Time	$\overline{T}_d$	$\overline{T}_i$	$\dfrac{d\overline{T}_d}{dt}$	A_s	q	h_d	h_c	E

$t =$

198 *J. Maddren and E. Marschall*

Notation

A_s surface area of drop
c_p heat capacity of drop
E heat-transfer efficiency
h heat-transfer coefficient
k thermal conductivity
q heat rate to drop
T temperature
T_{di} dispersed-phase nozzle inlet temperature
T_i interface temperature
t time
t_f time of drop formation
V volume of drop
z axial coordinate
ρ density of drop
τ thermocouple time constant

Subscripts:
c   continuous phase
d   dispersed phase

Fig. 25.4. Thermocouple probe assembly.

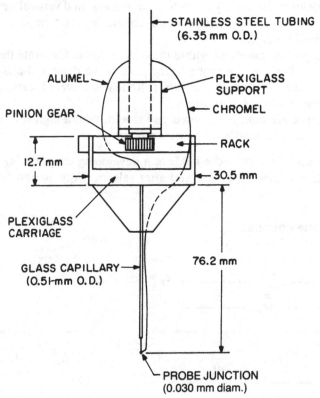

Fig. 25.5. Isotherms (°C) during drop formation and release (the solid line represents the liquid–liquid interface). Time of formation $t_f = 0.46$ seconds; efficiency at release $E = 16.6$ percent.

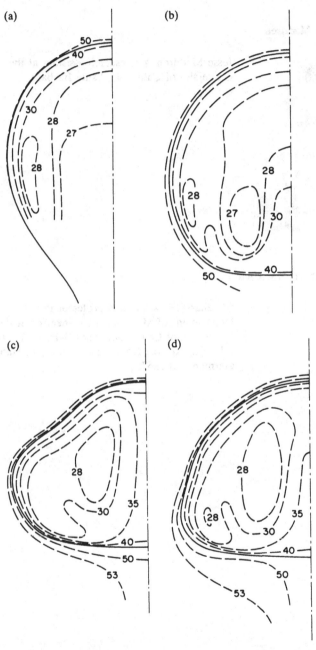

Superscripts:
$\bar{x}$ integrated average value of x

Jesse Maddren

Jesse Maddren is a research assistant at the University of California, Santa Barbara.

Ekkehard Marschall

Ekkehard Marschall is a professor in the Department of Mechanical Engineering at the University of California, Santa Barbara. He received a D. Eng. degree from the University of Hannover, Germany, in 1967.

PART II
Experiments in thermodynamics

Thermodynamics is one of the major branches of physics. It is concerned with the behavior of energy as affected by changes of temperature. In particular, thermodynamics explains the observed properties of matter at any temperature. In this connection, we might consider heat capacities, magnetic and electrical effects, phase transitions, and higher-order transitions (such as the Ehrenfest third-order transition) as principle topics.

Classical thermodynamics on the other hand treats the many observable properties of solids and fluids in such a manner that they can all be viewed as a consequence of a few. The four laws of thermodynamics are the result

Fig. II.1.1. Coherent structures in thermal turbulence. Swirls and plumes are injected from the top thermal boundary layer in turbulent Rayleigh–Bénard convection. Visualizations are made using thermochromic liquid crystals in water ($Pr = 6$) at $Ra = 10^9$. The cell is a cube of length 18.65 cm. Field of view is approximately 2×2 cm^2. (Courtesy of G. Zocchi, E. Moses, and A. Libchaber, *Phys. Fluids A* 3, 9 (1991): 2036.)

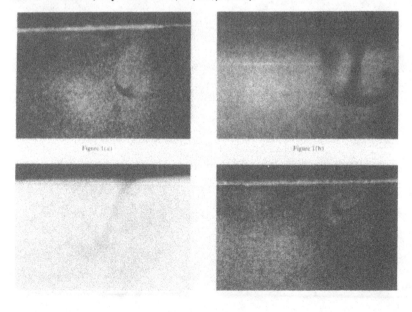

Fig. II.1.2. Dropwise condensation. In all of the photos, one observes the nuclei appearing on the bare cooling surface followed by a very thin film as density increases on which many speckles appear after the interference stripes disappear. Then the speckles fluctuate slightly. After a certain (critical) thickness is reached, the unstable film fractures into many tiny droplets. (Courtesy of S. Sugawara and K. Katsuta, *3rd International Heat Transfer Conf., Chicago, IL*, AICE, New York, 1966.)

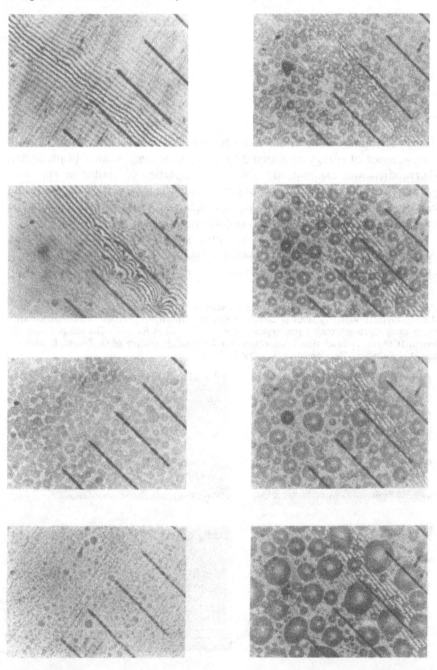

of observation: thus the importance of experimentation in this science. The development of the four laws is elegant. The laws contain an aesthetic spirit that once grasped and understood by the student will stand as the undercurrent for all the other physical sciences.

To tickle the student's imagination consider the application of thermodynamics to one aspect of the study of black holes. It is known a black hole has entropy. For example, the area of the event horizon of a black hole is entropy. Adding mass to a black hole increases the event horizon since it has added entropy. If the black hole has entropy it has temperature, which means black holes can radiate energy. The question arises how can black holes (possessing temperature) emit particles of radiant energy if nothing can escape past the event horizon?

Realizing the universe is not empty (i.e., it has no empty space and only a minimum amount of uncertainty regarding its composition), consider a pair of particles existing outside but near the event horizon of a black hole. Consider the particles "borrowing energy" from the energy of the universe (one borrowing positive energy and one borrowing negative energy so that the net energy cancels out). As the particles move, suppose the particle with the negative energy crosses the event horizon into a black hole, whereas the particle with the positive energy remains outside the event horizon as radiant energy. The negative energized particle must decrease the entropy, and hence the event horizon area decreases. Carrying this further, if enough negative energized particles enter a black hole, the black hole could disappear. If its energy is reduced so must be its mass (by Einstein's $E = mc^2$). The smaller a black hole, the hotter its surface temperature; so hot it could radiate like a star. As yet, we have no experiment to simulate this; but, according to some quantum cosmologists, such as Professor Alan Guth of M.I.T., it is not inconceivable that engineers of the future could create a universe in the laboratory.

On a much smaller scale, we present in this part the favorite experiments on thermodynamics from a few eminent scientists and engineers. Hopefully, they may help establish the student's understanding of thermodynamics so that he or she might be the one to tackle experiments on a cosmological scale.

EXPERIMENT 26

Effect of surface condition on attainable superheat of water

Contributed by
R. L. WEBB

Principle

A fluid must be superheated above its saturation temperature for nucleate boiling to occur. The amount of liquid superheat sustainable is influenced by the presence (or absence) of nucleation sites.

Object

To measure the superheat required to boil water for containers having different surface characteristics.

Background

At one atmosphere pressure, the saturation temperature of water is 100 °C. When a pan of water is heated to 100 °C on an electric heating element, you will observe columns of steam bubbles rising from the base surface of the container. This is nucleate boiling. However, the water in the thin thermal boundary layer near the heating surface is superheated to a temperature greater than 100 °C. Thus, its temperature is greater than the 100 °C saturation temperature.

This experiment demonstrates the degree to which water may be superheated, and the effect of "nucleation sites" in the container wall on the amount of superheating that occurs.

A microwave oven is used for the experiment, because the microwave supplies approximately uniform heat per unit volume to the water.

Apparatus

One litre of degassed water
A clear glass beaker

205

A microwave oven
A thermocouple to measure the water temperature
A handful of clean sand, or other insoluble granular material

Procedure

The water is first "degassed" by vigorously boiling it in a pan, using a surface heating unit.

The water is then poured in the glass beaker, and the thermocouple positioned at the center of the beaker. The beaker is then placed in a microwave oven, and full power is applied to the oven.*

Measure the water temperature as a function of time. After several minutes, you should observe a sudden conversion of water to steam. Write down the last observed water temperature. This liquid-to-vapor conversion should be accompanied by a noticeable sound like "whoomp!"

Now pour some sand into the container, and add enough water to replace that which you boiled off. Repeat the experiment. You should observe that you cannot heat the water to as high a temperature as in the previous experiment. This is because nucleation occurs in the thin bed of sand particles at the bottom of the beaker.

Questions

1. Is the maximum superheat attainable with the glass beaker influenced by use of distilled water, as opposed to tap water?
2. If the water is not degassed before running the experiment, will the answer change?
3. An unglazed, clay pot has a porous surface, which may provide nucleation sites. Rerun the experiment using this.

Suggested headings

Constants:

$p = $ _____ mm Hg; $T_{sat} = $ _____ °C

Beaker material	Particles added to water	Water condition	T	$T - T_{sat}$

IV (column label) · Observation (MV)

---

* Electrical arcing from the thermocouple will occur when heating the water in the microwave oven. A temperature sensor made of material that will not produce arcing must be used.

Ralph L. Webb

Ralph L. Webb is Professor of Mechanical Engineering at Pennsylvania State University. From 1963 to 1977 he was manager of heat-transfer research for the Trane Co., LaCrosse, Wisconsin. He received his Ph.D. from the University of Minnesota, has published in the area of heat-transfer augmentation, and holds three U.S. patents on enhanced heat-transfer surfaces. He teaches academic courses in "heat-exchanger design" and "enhanced heat transfer" at Penn State, and is currently performing research in boiling, condensation, fouling, heat pumps, and automobile radiators. Much of this research involves enhanced-heat-transfer surface technology.

EXPERIMENT 27
Experiments for compressibility and vapor pressure

Contributed by
R. A. GAGGIOLI *and* W. J. WEPFER

Principle

An experiment that involves the measurement of thermostatic properties is described. Temperatures and pressures of a real gas are measured in order that $v(T,P)$ can be constructed. In addition the saturation pressure as a function of temperature $P_s(T)$ is measured so that heats of vaporization can be evaluated.

Object

The primary objective of this experiment is the determination of $v(T,P)$ and $P_s(T)$ over a specified range of temperatures and pressures. In addition, other goals include the: (1) comparison of the measured $v(T,P)$ data with published data and with $v(T,P)$ functions obtained from the principle of corresponding states, and (2) comparison of the measured $P_s(T)$ data with published data and the approximation $\ln (P_s(T)) = m/T + b$.

Background

Most processes, devices, and systems built by engineers utilize gases and liquids. Such analysis and design invariably requires the evaluation of various thermostatic properties. For gases all thermostatic properties can be determined from knowledge of the "mechanical" equation of state $v(T,P)$ and either the "thermal or caloric" equation of state $(U = f(T,P))$ or $c_p(T)$, which is the perfect-gas heat capacity. In addition to these functions, knowledge of $P_s(T)$ provides a means of evaluating the thermostatic properties for two-phase materials. Finally, excursions into the liquid region also require that the heat capacity of the liquid be known.

Apparatus

The basic equipment required to conduct these measurements includes sample cylinders, pressure-measurement and -calibration hardware, balance,

208

temperature-controlled baths, vacuum, pump, and appropriate fittings. This experiment is easily partitioned into several "stations" wherein the students perform, in sequence, each of their assigned tasks. For a small-sized lecture sample cylinders of a known volume are most appropriate (Fig. 27.1).

A small safety relief valve must be installed at the bottom of the sample cylinder. The results presented subsequenty for the saturation pressure of R-12 were obtained from cylinders having $V = 75$ ml.

The cylinders are instrumented with Bourdon-tube-type pressure gauges. However, such gauges must be calibrated. In most cases a "calibration station" consisting of a simple deadweight tester will suffice.

This experiment requires that the cylinders be evacuated. It is most convenient to construct an "evacuation station" consisting of a vacuum pump and associated fittings and quick disconnects so that the students can quickly purge their cylinders of air.

The next step is carried out at the "filling station," which is comprised of the source cylinder containing the gas or material to be tested and the associated fittings and quick disconnects.

A "weighing station" consisting of an electronic scale is required.

Several temperature-controlled baths containing a sufficient quantity of water at temperature intervals spanning the range of interest must be set up.

Fig. 27.1. 75-ml sample cylinder with inlet valving, pressure gauge, and safety relief valve (bottom).

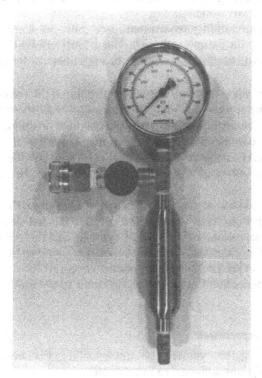

The number and type of baths is a function of the temperature range over which the measurements are required as well as a function of the funds available. Moderately priced refrigerated baths extend the measurement range down to −20 °C.

Finally, a station for the safe discharge of the sample gas is required.

Depending on the number of students that are in the laboratory, the cost of the fittings is typically less than $100.

A 75 ml sample cylinder with a safety relief valve runs $75.00.

All remaining equipment is general laboratory hardware that is usually available.

Procedure

The initial step of this experiment is the calibration of the pressure gauge that is mounted on the sample cylinder.

The next step is the evacuation of the cylinder.

The cylinder is connected to the vacuum pump at the "evacuation station." Care must be exercised to open and close the various valves in the proper sequence to prevent air from reentering the cylinder.

It is helpful to use a blow dryer to heat the cylinder during evacuation. This serves to liberate any condensables from the interior walls of the sample cylinder. At this point the cylinder is closed, its exterior is wiped dry, and it is weighed (this gives the tare weight).

In the case of the compressibility experiment, care must be exercised to insure that only gas enters the sample cylinder. This is fairly straightforward due to the pressure drop of the fluid as it flows through lines to the sample cylinder.

On the other hand, the vapor-pressure experiment requires the transfer of a two-phase mixture to the sample cylinder. In this case, the source cylinder containing the material to be tested should either have a draw or siphon tube.

Another option is the commercially available canister used to fill residential or automotive HVAC systems. These small canisters provide liquid when turned upside down but supply gas when discharged from the upright position.

In both cases, it is helpful to cool the sample cylinder in an ice bath before filling it with the liquid–gas mixture. The amount of material that should be transferred depends on the volume of the sample cylinder and the material's thermostatic properties. For example, roughly 30 g of R-12 need to be transferred to a 75-ml sample cylinder in order to obtain good results.

Once the sample cylinder is filled, it is weighed in order to obtain the mass of material to be tested.

Compressibility experiment

At the first temperature T_1, the cylinder containing only gas is submersed into the water bath (at T_1) and allowed to equilibrate.

Upon reaching equilibrium, the temperature T_1 and pressure are recorded. The cylinder valve is now opened to let a small amount of the gas escape. The cylinder is weighed and resubmersed into the same water bath (at T_1), and upon reaching equilibrium, the pressure and temperature are recorded.

This procedure is repeated at four to six additional pressures down to atmospheric.

If it is desired to generate $v(T,P)$ at another temperature T_n, the procedure is repeated in a water bath maintained at T_n.

Vapor-pressure experiment

These measurements are somewhat simpler since the mass is kept constant.

The sample cylinder containing the two-phase mixture is submersed into a water bath, is allowed to equilibrate, and has its temperature and pressure recorded.

This is repeated in several different water baths, each maintained at a different temperature.

It is important that the students carefully watch the pressure gauge in order to prevent the cylinder from quickly rising above the critical pressure.

It is advisable, as a safety precaution, to have an extra cold bath available for dunking overheated cylinders.

Results

Figure 27.2 shows a plot of the compressibility factor $Z(T_r,P_r)$ as a function of P_r for $T_r = 0.989$ (0 °C) for CO_2 as obtained from experimental measurements. The experimental results are given along with data given by ASHRAE[1] and data obtained from the compressibility chart.[3]

Table 27.1 and Fig. 27.3 show the results of vapor-pressure measurements made for R-12. Note the good agreement for saturation pressure between the experimental values and standard R-12 data as found in most undergraduate thermodynamics texts.[3] The enthalpy of vaporization h_{fg} is equal to m/R, where m is the coefficient of $1/T$ in the regression equation for $\ln(P_s(T))$ and R is the gas constant for R-12. Table 27.1 gives values of h_{fg} that were obtained from a series of piecewise linear regressions for $\ln(P_s(T))$. Agreement with tabulated values for h_{fg} is quite good considering the simplicity of the experiment. Table 27.1 is easily programmed as a spreadsheet which enables the students to assess their data as they proceed with the experiment.

Questions

1. Use the experimentally determined $v(T,P)$ to calculate the real gas enthalpy deviation, $h - h^*$, for your material at one of the temperatures at which you

made measurements. Compare this value with the value obtained from the generalized enthalpy-deviation chart.

2. Compare the experimentally determined $v(T,P)$ with one of the well-known equations of state.

3. How does the presence of a small amount of air in the cylinder affect (a) the $v(T,P)$ data? (b) the vapor-pressure data?

4. Over a small range of temperatures (e.g., 20 °C), use the experimental vapor-pressure data to obtain the coefficients for the Antoine vapor-pressure equation

$$\ln(P_s) = A - \frac{B}{T + C} \tag{27.1}$$

5. Using the coefficient of thermal expansion for the sample cylinder (i.e., for stainless steel), estimate the error introduced due to expansion of the sample cylinder.

Conclusions

These experiments provide the undergraduate engineering student with an excellent background in the measurement and use of thermostatic properties. The experiments are inexpensive and easy to run, and the data is quite

Fig. 27.2. Compressibility factor of CO_2 plotted as a function of reduced pressure for the isotherm $T_r = 0.898$ (0 °C). Tabular data obtained from ASHRAE.[1] Compressibility data obtained from Obert.[3]

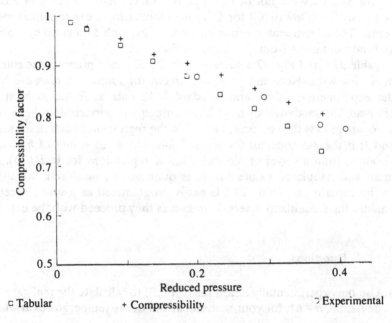

□ Tabular   + Compressibility   ɔ Experimental

Table 27.1. *Vapor pressure data for r-12*

CYLINDER	3
AMBIENT TEMPERATURE (C):	20
BAROMETRIC PRESSURE (mmHg):	747.3
CYLINDER VOLUME (cm³):	75

MASS OF EVACUATED CYLINDER (kg):	1051.3
MASS OF FILLED CYLINDER (kg):	1084.1
MASS OF R-12 (kg):	32.8

TEMP (°C)	TEMP (K)	1/T (1/K)	MSRD GAUGE PRES (kPa)	CORRTD GAUGE PRES (kPa)	ABS PRES (kPa)	PRES PROP (TABLES) (kPa)	ln (PRES) (EXPTL)	ln (PRES) (TABLES)	ln (PRES) REGRESS
3.0	276.15	3.62E-03	200	204.6	304.2	341.0	5.718	5.832	5.957
15.0	288.15	3.47E-03	500	511.5	611.1	491.4	6.415	6.197	6.258
30.0	303.15	3.31E-03	675	690.5	790.1	744.9	6.672	6.613	6.579
45.0	318.15	3.14E-03	1040	1063.9	1163.5	1084.3	7.059	6.989	6.920
60.0	333.15	2.87E-03	1400	1432.2	1531.8	1514.9	7.334	7.323	7.462
75.0	348.15	2.75E-03	2000	2046.1	2145.7	2064.3	7.671	7.633	7.703

TEMP (°C)	LIQUID SPEC VOLUME (m³/kg)	VAPOR SPEC VOLUME (m³/kg)	SPEC VOL V_{fg} (m³/kg)	EXPTL SPEC VOL (m³/kg)	QUAL	(EXPTL) ENTHALPY H_{fg} (kJ/kg)	(TABULAR) ENTHALPY H_{fg} (kJ/kg)
3.0	7.21E-04	5.06E-02	4.99E-02	2.29E-03	0.0314		143.7
15.0	7.43E-04	3.45E-02	3.47E-02	2.29E-03	0.0445	210.6	135.1
30.0	7.74E-04	2.35E-02	2.27E-02	2.29E-03	0.0665	134.4	128.1
45.0	8.11E-04	1.60E-02	1.52E-02	2.29E-03	0.0969	100.4	113.5
60.0	8.58E-04	1.11E-02	1.03E-02	2.29E-03	0.1393	101.5	
75.0	9.20E-04	7.72E-03	6.80E-30	2.29E-03	0.2009		

reproducible. Using single stations for "calibrating," "evacuating," "filling," and "weighing" and with multiple temperature-controlled baths, groups of four students can easily complete both experiments within a three-hour laboratory period.

Acknowledgments

The authors express their appreciation to Dr. C. W. Savery (formerly of Drexel University) for many helpful suggestions. In addition, we thank the Westinghouse Educational Foundation and the Georgia Tech Foundation which supported the development of this experiment at Marquette University and Georgia Tech, respectively. Laboratory manual write-ups of these experiments as currently used at Georgia Tech and Marquette are available from either author (W. J. Wepfer is at the George W. Woodruff School of Mechanical Engineering, Georgia Institute of Technology, Atlanta, GA 30332-0405; R. A. Gaggioli is at the University of Lowell, Lowell, MA 01854).

Suggested headings

Constants: Ambient temperature: _____ ; Mass of evacuated cylinder: _____

Barometric pressure: _____ ; Mass of filled cylinder: _____

Cylinder volume: _____ ; Mass of sample material: _____

Fig. 27.3. Vapor-pressure curve for R-12. Pressure is given in kPa. Tabular data obtained from ASHRAE.[1]

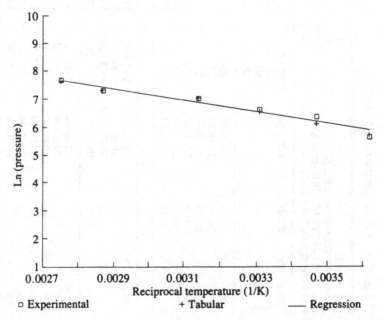

T(°C)	T(K)	1/T	Gauge pressure (msrd)	Corrected gauge pressure	Absolute pressure	Pressure property pressure	ln (pressure) (exptl)	ln (pressure) (tables)	ln (pressure) regression

T(°C)	Liquid specific volume (tables)	Vapor specific volume (tables)	V_{fg} (tables)	Specific volume (exptl)	Quality (exptl)	Enthalpy h_{fg} (exptl)	Enthalpy h_{fg} (tabular)

References

1. *Thermodynamic Properties of Refrigerants*, ASHRAE, Atlanta, GA, 1980.
2. Gaggioli, R. A., and Wepfer, W. J., "Instructional experiments for compressibility and vapour pressure," *Intl. J. Mech. Engrg. Ed.*, 9, 2 (1981): 111–21.
3. Obert, E. F., and Gaggioli, R. A., *Thermodynamics*, 2nd ed., McGraw-Hill, New York, 1963.

William J. Wepfer

William J. Wepfer is Associate Professor of Mechanical Engineering at Georgia Institute of Technology. He received his Ph.D. in 1979 from the University of Wisconsin. Has served as a consultant to Professional Engineering Consultants, Milliken & Company, Wave Air Corporation, and the Teltech Resource Network. Dr. Wepfer has supervised and participated in grants and contracts from the NSF, Engineering Foundation, ASHRAE, USAF, Lockheed-Georgia, and Georgia Power Company. Professor Wepfer's research interests span from thermal systems analysis and development to applied heat transfer. Dr. Wepfer is lead author of the forthcoming text *Engineering Measurement and Instrumentation* and the recipient of numerous awards.

Richard A. Gaggioli

Richard A. Gaggioli has been Professor of Mechanical and Energy Engineering at the University of Lowell since 1985. He received his Ph.D. in 1961 from the University of Wisconsin. His positions have included Research Member at U.S. Army Math. Research Center and Associate Professor of Mechanical Engineering at University of Wisconsin, 1961–9, and Professor at Marquette University, 1969–81. Dr. Gaggioli's publications span from theoretical to applied thermodynamics and mathematics; he is co-author of a textbook, *Thermodynamics*, and editor of several research volumes. He has worked on such topics as modeling, optimization of energy and chemical plants, and laboratory development.

EXPERIMENT 28
Determination of time constants

Contributed by
G. P. BAL

Principle

The time constant of a thermometer (or a thermocouple) is a measure of the speed of response. It represents the time required to complete 63.2 percent of the total change.

Object

The objective of this experiment is to determine the time constants of a thermocouple and a thermometer by graphical analysis of their temperature–time curves.

Background

Most temperature-measuring devices (thermocouples, thermometers, etc.) respond as first-order systems. When subjected to a sudden (step) change in temperature, an exponential type of response occurs. Energy balance on the sensing element of the device results in the following differential equation:

$$Mc\frac{dT}{dt} + hA(T - T_\infty) = 0 \tag{28.1}$$

The solution of Eq. (28.1) to a step input gives the temperature of the sensor as a function of time:

$$\frac{T_\infty - T}{T_\infty - T_0} = e^{-t/\tau} \tag{28.2}$$

where $\tau = Mc/hA$. τ is called the time constant and is usually expressed in seconds.

Equation (28.2) may be rearranged in a more convenient form by taking logarithms on both sides and introducing the dimensionless variable

217

$$\theta = \frac{T_\infty - T}{T_\infty - T_o} \tag{28.4}$$

Then

$$\ln(\theta) = -t/\tau$$

or

$$\ln(\theta) = mt \tag{28.5}$$

where m is the slope of the $\ln(\theta)$ = versus-t line. From Eq. (28.5) it can be observed that the time constant is the negative reciprocal of the slope m.

Equation (28.2) shows that the speed of response depends only on the value of τ. If the magnitude of one time constant is substituted for t in Eq. (28.2), it is seen that 63.2 percent of the total change will have occurred. In practice, it is assumed that a response is completed during a period of five time constants.

Apparatus

Stirring hot plate with beaker of distilled water
Mercury-in-glass thermometer and stopwatch
Storage oscilloscope
Thermocouples (about 3 to 4 ft. long)

The apparatus is shown in Fig. 28.1.

Procedure

The time constant was determined by graphical analysis of the temperature-versus-time curve. The instrumentation and the procedure to produce an

Fig. 28.1. Thermocouple time-constant apparatus.

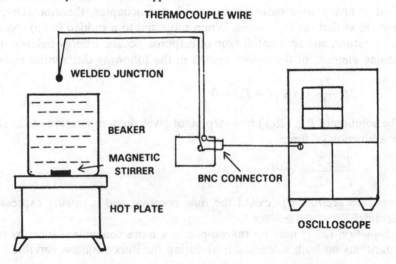

THERMOCOUPLE WIRE

WELDED JUNCTION

BEAKER

MAGNETIC STIRRER

BNC CONNECTOR

HOT PLATE

OSCILLOSCOPE

accurate temperature-versus-time (*T–t*) curve is different for thermocouples and thermometers.

Thermocouple: The procedure described uses a storage oscilloscope. Connect the thermocouple directly to the oscilloscope. Select the mV/div scale such that the displayed temperature change covers about 6 divisions. Select single-sweep operation. Select STORAGE mode.

With thermocouple junction at room temperature, produce the initial temperature trace on the CRT by depressing and releasing the single-sweep knob.

Next, produce the *T–t* curve by carefully coordinating the release of the single-sweep knob and the immersion of the thermocouple junction into the boiling water. This may require several tries and different TIME/DIV settings. For an average thermocouple junction 50 msec/div gives a good resolution on the time axis.

It is important to transfer the thermocouple into water quickly to ensure a true step input. Hold the the thermocouple in water for a sufficient period of time (depending on the junction size) until the response is essentially complete.

Obtain the trace of the final temperature by depressing and releasing the single-sweep knob.

Transfer the CRT trace onto a clear plastic overlay with matching oscilloscope grid for graphical analysis.

Thermometer: For the thermometer–stopwatch combination, read and record the room temperature.

Immerse the thermometer in the boiling water and hold it there until the reading is stable.

Remove the thermometer from the water quickly and start the stopwatch simultaneously.

Record temperature–time readings every second for the first few seconds and then at increased time intervals until the thermometer reading approaches the room temperature.

Synchronization between the thermometer reading and the stopwatch reading is difficult in the initial few seconds of the experiment.

The collection of data requires coordination between the thermometer reader, the stopwatch reader, and the data recorder.

Plot the temperature-versus-time data on semilog paper for analysis.

Results

Analyze the T–t curves to determine the time constant. For the thermocouple experiment, mark 63.2 percent of the total change in temperature. Determine the corresponding time, which is the time constant. (See Fig. 28.2.)

For the thermometer experiment calculate the slope of the best-fit line through the data points. The negative reciprocal of the slope should be the time constant. (See Fig. 28.3.)

It is worthwhile to point out that because the time constant depends on the convective heat-transfer coefficient, one cannot specify the time constant without specifying the fluid and the heat-transfer conditions (e.g., free or forced convection). For example, a thermometer in gently stirred water might have a time constant of 10 seconds whereas the same thermometer in stagnant air would possibly have a time constant of 60 seconds.

Suggested headings

Constants: $T_{\infty} =$ _____ ; $T_0 =$ _____

IV	MV
Time (sec)	Temperature (°C)

Reference

1. Bal, G. P., and Schiller, R. W., *Notes on Instrumentation*, Pennsylvania State University, Harrisburg, PA, 1987.

Fig. 28.2. Simulated oscilloscope temp–time trace.

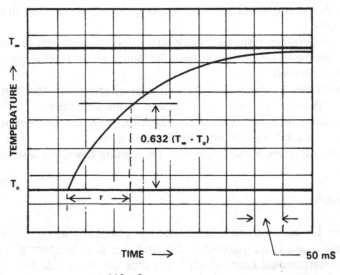

$\tau = 110 \text{ mS}$

Notation

A	surface area
c	specific heat
h	convective heat-transfer coefficient
M	mass
m	slope
t	time
T	temperature of the sensor
T_∞	fluid temperature
T_0	initial temperature
τ	time constant
θ	temperature ratio

Fig. 28.3. Temperature response of a thermometer.

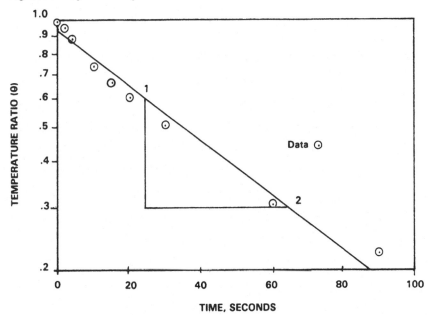

$$\text{Time Constant } (\tau) = -\frac{1}{m}$$

$$\text{where } m = \text{slope} = \frac{\ln\theta_2 - \ln\theta_1}{t_2 - t_1} = -0.01717/s$$

$$\tau \approx 58 \text{ sec}$$

Ganesh P. Bal

Ganesh P. Bal is Assistant Professor of Engineering at Pennsylvania State University, Harrisburg. Dr. Bal received his Ph.D. degree in mechanical engineering from Penn State University in 1980. His specialities include heat transfer, instrumentation, and finite-element analysis.

EXPERIMENT 29

Very fast versus very slow processes: Which are more efficient (closer to reversibility)?

Contributed by
ELIAS P. GYFTOPOULOS, MICHAEL STOUKIDES,
and MIGUEL MENDEZ

Principle

Transfer of electricity out of a storage battery is much more efficient (closer to reversibility) when it is very fast rather than very slow.

Object

It is often argued that reversible processes take an infinite time to complete and, therefore, are of questionable usefulness. Although there is truth in this argument for certain processes, such as transfer of energy across a finite temperature difference, the argument is neither universally valid nor representative of some practical phenomena. A simple and very important counterexample is provided by a storage battery. If discharged quickly, the battery does work almost equal to the stored energy. If discharged more slowly than the rate of its internal discharge (let alone infinitely more slowly), the battery does practically no work, that is, all its availability is dissipated. The availability is dissipated because the internal discharge generates entropy spontaneously or, said differently, the internal discharge is irreversible.

Apparatus

As shown schematically in Fig. 29.1, the apparatus consists of a cell with two electrodes (1), a temperature bath (2), a glass thermometer and an electric heating plate (3), a galvanostat (4), a resistance box (5), an ammeter (6), and a voltmeter (7). Another schematic of the apparatus is shown in Fig. 29.2.

The cell consists of a 3-liter Pyrex beaker that contains two liters of sulfuric-acid aqueous electrolyte and two lead electrodes. The density of H_2SO_4 is 1.28 g/cm^3. The H_2SO_4 solution may be prepared by diluting 660 cm^3 of H_2SO_4 in 1340 cm^3 of water.

Each electrode is made of 5.5-cm-x-6.2-cm-x-0.038-mm (1.5-ml) lead foil. Each foil is sandwiched between two 17-cm-x-7-cm-x-0.6-cm plexiglass frames with a 5-cm-x-5-cm window at the location of the foil (see Fig. 29.3). Thus, each foil can be held firmly at the desired position. The distance between the electrodes is about 7 cm.

The cell is immersed in a 5-liter beaker filled with water or oil which serves as a temperature bath. The temperature of the bath is controlled by the electric heating plate, and measured by the glass thermometer. Temperatures in the range 20 °C to 60 °C are used.

The galvanostat is model AMEL #549, and supplies direct current up to 1000 mA. The resistance box provides a resistive load in the range from 1 to 10^5 ohms. An ammeter (1 μA to 2 A) and a voltmeter (1 mV to 1 kV), or two multimeters, are used for current and voltage measurements.

Procedure

The cell is charged by a current $I = 750$ mA for 30 minutes. During the charging period the voltmeter reads between 3 and 3.5 volts, and the cell temperature is kept at 25 °C.

Two procedures are used for the discharge experiments. In the first, the work-producing procedure (WP), the cell is discharged through a resistive load in the range between 200 and 2000 ohms.

Fig. 29.1. Schematic of experimental apparatus: (1) cell; (2) water bath; (3) heating plate; (4) galvanostat; (5) resistance box; (6) ammeter; (7) voltmeter.

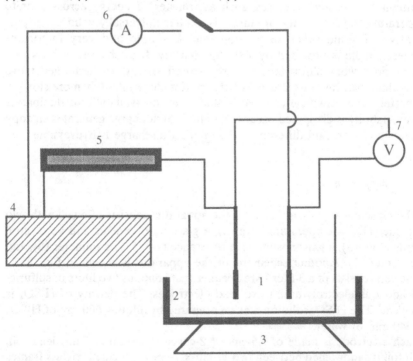

The current I through the load and the voltage V across the load are recorded as functions of time, beginning at the instant the charging phase is interrupted, and lasting until the cell is dead.

A graph of the product $P = I \times V$ versus time is made. It represents the power or work per unit time done by the cell. The integral under the power graph is the work done by the cell.

In the second, the self-discharge procedure (SD), no resistive load is connected at the end of the charging phase, and no work is done by the cell. Only from time to time is the voltage across the cell observed by connecting the electrodes to a large resistor (3000 ohm or larger) for a few seconds. Each self-discharge experiment lasts several hours, until the cell is dead.

Fig. 29.2. Schematic of temperature-controlled electrochemical cell: (1) plexiglass frame; (2) lead contact; (3) lead electrode; (4) heat plate; (5) water bath; (6) sulfuric acid solution; (7) glass container.

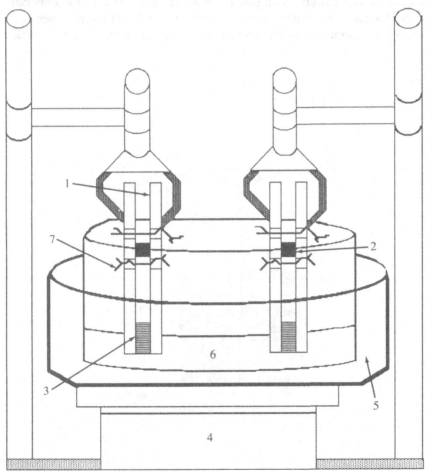

Results

Typical experimental results are listed in Table 29.1. The table includes the type of experiment, the temperature of the cell, the resistive load, the length of time of the experiment, and the work done by the cell.

It is clear from these results that, if work is done at a rate faster than that dictated by the rate of self-discharge, the cell does relatively a lot of work. On the other hand, if no work is done (work is done at an almost zero rate), the self-discharge dissipates all the ability of the cell to do work, and no work can be transferred to an external load.

Of course, the reason for the dissipation is the spontaneous generation of entropy in the course of self-discharge, that is, irreversibility.

The data also show that, as temperature increases, the rate of self-discharge increases. The reason is that as temperature increases, internal reaction rates increase.[1]

More fundamentally, it is interesting to compare the transfer of energy out of a fixed-volume battery to an electric load, and the transfer of energy out of a fixed-volume, fixed-temperature reservoir to a lower-temperature reservoir. In the former case, the transfer is out of a state that is not a stable

Fig. 29.3. Front (a) and side (b) views of the electrodes.

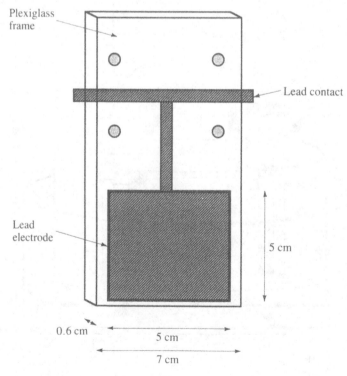

(a)

Table 29.1

Experiment type	Temperature °C	Resistance ohms	t^* hours	W^* joules
	24	400	2.5	59
	25	200	4.0	69
	25	1,000	4.5	45
Work-prodicing	25	2,000	6.0	40
	25	2,000	6.5	43
	42	200	0.9	185
	42	600	1.75	45
	45	2,000	3.3	25
	46	400	1.5	48
	55	3,000	1.9	11.4
	25	—	15.2	—
Self-discharge	43	—	4.3	—
	50	—	4.0	—
	55	—	2.3	—

Note: t^* = time required for the cell voltage to drop down to 3% of its final value.
W^* = Work produced from time zero to time t^*.

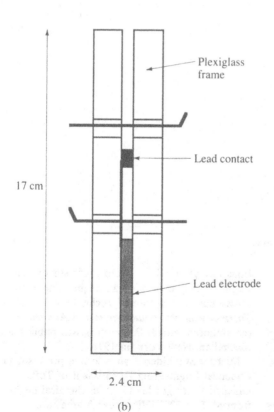

(b)

equilibrium state (not a thermodynamic equilibrium state). As such, the energy transfer need not be accompanied by entropy transfer and, therefore, no entropy need be accommodated in another system, such as a low-temperature reservoir. So the process can be fast without being irreversible. In the latter case, the energy transfer out of the high-temperature reservoir is unavoidably accompanied by entropy transfer. This entropy must be accommodated in the lower-temperature reservoir, and this accommodation unavoidably requires a generation of entropy by irreversibility. Moreover, for a given temperature difference, the rate of entropy generation reduces to zero only when the rate of energy transfer approaches zero or, equivalently, the process lasts infinitely long.

Suggested headings

WP experiment: $R =$ _____ ohms

Time t	Voltage V	Current I	Power $I \times V$	Work $W = \int_0^t P dt$
(minutes)	(volts)	(amperes)	(watts)	(joules)

SD experiment:

Time t	Voltage V
(minutes)	(volts)

Reference

1. Bode, H., *Lead Acid Batteries*, Wiley, New York, 1977.

Elias P. Gyftopoulos

Elias P. Gyftopoulos is Ford Professor of Mechanical Engineering and of Nuclear Engineering at the Massachusetts Institute of Technology. His book, *Thermodynamics: Foundations and Applications* (co-authored with J. P. Beretta), was published by Macmillan, New York, in 1991.

Michael Stoukides is an associate professor in the Chemical Engineering Department of Tufts University. He got his Ph.D. in chemical engineering from MIT in 1982. His research areas are heterogeneous catalysis, chemical-reaction engineering, and fuel cells.

Miguel Mendez was an undergraduate at Tufts University, and he got his B.S. in chemical engineering in 1992.

EXPERIMENT 30
Determination of volumetric fraction of each phase in multiphase flow

Contributed by
T. SAKAGUCHI

Principle

The flow of a mixture composed of more than two phases or two different kinds of immiscible liquids having different physical properties is referred to as a multiphase flow. A gas–liquid two-phase flow, a liquid–solid two-phase flow, a gas–liquid–solid three-phase flow, and a liquid–liquid two-phase flow are representative examples of multiphase flow. The density of the multiphase mixture is one of the more important variables to be evaluated when we apply the equations of continuity and of motion to multiphase flow to obtain such flow characteristics as pressure drops, velocities, and shear stresses. The density ρ of the multiphase mixture is expressed by Eq. (30.1) in terms of the density and volumetric fraction α of each phase.

$$\rho = \Sigma \alpha_i \rho_i \tag{30.1}$$

where the subscript i denotes G(gas), L(liquid), or S(solid). In order to estimate ρ by Eq. (30.1), we must develop a constitutive equation for the volumetric fraction of each phase. The constitutive equation for the volumetric fraction is obtained from experimental data. The volumetric fraction of each phase is measured by "a quick-closing-valve method."

The volume flow fraction β_i of phase i is an important fundamental variable defined by

$$\beta_i = \frac{Q_i}{Q_T} \tag{30.2}$$

where Q_i is the volume flow rate of phase i, and Q_T is the total volume flow rate. This volume flow fraction is not used to calculate the density of the multiphase mixture since there are slips between phases. Except for the case of no slip, it does not correspond to the actual volumetric fraction of phase i in the pipe.

Object

The objective of this experiment is to determine quantitatively the volumetric fractions of each phase of the multiphase mixture flowing in a pipe by the

230

quick-closing-valve method. The volumetric fraction α_i of phase i is defined by the ratio of the volume V_i occupied by the i phase in a control volume to the total volume V_T of the control volume:

$$\alpha_i = \frac{V_i}{V_T} \tag{30.3}$$

Each volume is measured by the height of each phase trapped in a vertical pipe by quickly closing the valves that have been installed on the pipe. In the case of a horizontal pipe or an inclined pipe, each volume trapped in the pipe can not be measured by its height. Then, the trapped substances are taken out from the pipe and their volumes are measured by their heights in a graduated cylinder or their weights.

Background

Each phase in multiphase flow is not necessarily mixed homogeneously like a solution but flows separately with interfaces in a pipe. Each phase has radial (r), circumferential (θ), and longitudinal (z) distributions in a pipe. The distribution patterns are unsteady. In other words, a kind of phase existing at a local point changes with time (t).

As expressed in Eq. (30.3), the volumetric fraction α_i of the i phase is the ratio of the volume V_i occupied momentarily by the i phase in the fixed control volume to the total volume V_T of the fixed control volume.

The same process can be applied to a fixed control area and to a fixed control line. Then, an area-averaged and a line-averaged volumetric fraction of the i phase can be defined as follows:

$$\left\langle \alpha_i \right\rangle_A = \frac{A_i}{A_T} \tag{30.4}$$

$$\left\langle \alpha_i \right\rangle_L = \frac{L_i}{L_T} \tag{30.5}$$

Here A_T is the total area of the fixed control area, A_i is the area occupied by the i phase, L_T is the total length of the fixed control line, and L_i is the sum of the lengths of lines occupied by the i phase. These quantities are functions of time.

If one sums up residence times of each phase in a fixed space, a following time-averaged volumetric fraction of phase i can be defined as the ratio of the total residence time T_i of the i phase to the total time interval T_T

$$\left\langle \alpha_i \right\rangle_T = \frac{T_i}{T_T} \tag{30.6}$$

The term "volumetric fraction" is used here as the representative for various kinds of ratios of the phase in the space and in the time.

Many kinds of measuring techniques have been developed for each of the afore mentioned values. The quick-closing-valve method has been developed to initially measure the volumetric fraction. One experimental datum is obtained by one closure of valves. The time- and chord-averaged value of volumetric fraction was measured next by the attenuation method of a radioactive ray or X-ray through the multiphase flow. By using several measured values of attenuated rays through several chords at a fixed longitudinal position, the time- and area-averaged values can be calculated. Electric probes, optical probes, and isokinematic sampling probes have been developed to measure the time-averaged value of volumetric fraction at a fixed point. The change of phase at a fixed point is measured and recorded by the electric or the optical probe. The recorded wave consists of the combination of residence time periods of each phase at the tip of each probe. The time-averaged values of each volumetric fraction at the fixed point can be calculated as the ratio of time periods of each phase to a total measuring time T_T

$$\left\langle \alpha_i \right\rangle_P = \left(\frac{T_i}{T_T} \right)_P \tag{30.7}$$

where T_i is the sum of the residence time at the fixed point of phase i. By traversing this probe radially, the radial distribution of time-averaged volumetric fraction can be obtained. The time- and area-averaged values of volumetric fraction can be calculated by integrating their radial distribution assuming that the flow is circumferentially symmetric about the flow axis. Recently, the three-dimensional distribution of each phase was measured by image processing of data recorded by a high-speed video tape recording system. Please refer to the references 1, 2 for the other methods described here, if you want to obtain detailed information about them.

The quick-closing-valve method is particularly the more fundamental and the standard measuring technique as compared against other methods. Values of the flow parameters obtained by the quick-closing-valve method are fundamental and are used as standard data in calibration as well as to examine measured values obtained using such other methods as attenuation of rays of light, radioactive rays, and ultrasonic waves through the multiphase flow, the change of electric capacitance caused by the multiphase flow, and so on.

Apparatus

The experimental apparatus is shown schematically in Fig. 30.1. This experimental apparatus can be used to conduct experiments in a gas–liquid two-phase flow, a liquid–solid two-phase flow, and a gas–liquid–solid three-phase flow.

In the gas phase, the gas is supplied by a compressor.

For the liquid and solid phase, the substances are fed by a Mohno pump (such as a screw pump).

Flow rates should be kept at desired values during the measurement of the flow characteristics of multiphase flow.

Fig. 30.1. Experimental apparatus.

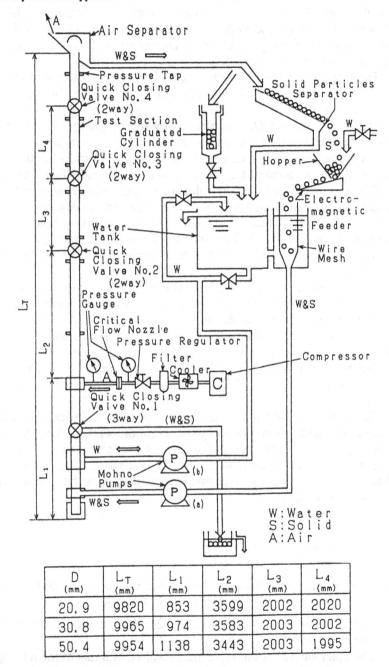

D (mm)	L$_T$ (mm)	L$_1$ (mm)	L$_2$ (mm)	L$_3$ (mm)	L$_4$ (mm)
20.9	9820	853	3599	2002	2020
30.8	9965	974	3583	2003	2002
50.4	9954	1138	3443	2003	1995

The critical-flow phenomenon through a nozzle is used to keep the inlet gas flow rate Q_G constant. In this case, the gas velocity through the critical-flow nozzle is determined by the pressure P_u at the upstream end of the critical-flow nozzle. Thus, the gas flow rate is controlled by the value of the pressure which is regulated by the regulator.

The relationship between flow rate and pressure is calibrated in advance.

The liquid flow rate Q_L is controlled by the revolutions of a high-pressure screw pump.

The solid flow rate Q_s is controlled by the frequency of vibration of an electromagnetic feeder and its throttle gate valve to a hopper. They are measured by measuring a time interval T and the volume of liquid and solid phase, v_L and v_s, accumulated in the graduated cylinder shown in Fig. 30.1.

The relationship between the solid volumes and their heights in the graduated cylinder is calibrated for each kind of particle in advance.

The values of Q_L and Q_s are obtained by dividing v_L and v_s by T, respectively. The volumetric fluxes of each phase, J_G, J_L, and J_s, are calculated by dividing Q_G, Q_L, and Q_s by the cross-sectional area of the test pipe, respectively.

Since the actual volume of each phase in the pipe is not necessarily equal to the volume flow fraction, it must be measured in situ by trapping the mixture instantaneously in the pipe.

Four valves are installed in the present experimental apparatus. Valve 1 is a three-way valve and the other three valves are two-way valves. Their schematics are shown in Fig. 30.2. These four valves must be closed instantaneously and simultaneously. To achieve this, solenoidal valves or a connecting rod are used.

The volume of each phase trapped in the space between valves 2 and 3, and valves 3 and 4, is measured by the height of its accumulated particulates. The relationship between height and volume is calibrated in advance.

Examples of the volumetric fraction of each phase in the gas–liquid–solid three-phase flow under the specific experimental conditions given in Table 30.1 are shown in Fig. 30.3.

Procedure

It is important to mention a number of key points and suggestions to make the experiment run smoothly. They are as follows:

1. Ball valves or cocks with the same inner diameter as the test pipe are used frequently for shutoff valves.
2. The inner surface of the valve must connect with that of the test pipe with negligible difference and clearance.
3. The mixture supplied to the pipe from the mixing section is exhausted through the three-way valve, when the valves are closed. The three-way

valve is installed at the position near the downstream side of the mixing section. The distance between the three-way valve and the measuring position must be long enough so that the disturbance caused by the concavity of the three-way valve may settle.

4. One should pay particular attention to any liquid leak through the valves when the valves are closed.

5. The volumes trapped in the space between shutoff valves are measured by their heights. Calibration curves should be prepared in advance by using the actual test pipe including the valves, the actual fluid, and solid particles.

   The relationship between liquid height H_L and liquid volume V_L in the pipe is determined by the measurement of the height caused by pouring a known volume of water into the test pipe on the valve which was closed. By this process, the precise value of the volume of the space is obtained, even though the shape of the space near the valve is complicated and its volume is not easily measured.

   The same procedure is applied to the other side of the test pipe. Then, the total volume V_T between two shutoff valves is determined.

6. The relationship between solid height H_s and the volume of solid particles, V_s in the pipe is determined by the measurement of the height caused by placing a known volume of solid particles in the test pipe on the valve. At this time, the height of solid particles changes according to their settling condition in the test pipe. Identical settling conditions of solid particles can be achieved by vibrating the pipe.

7. When the solid particles are submersed in the liquid phase, the volume of liquid phase is obtained by the difference of the volume calculated from the liquid height and the volume of solid particles.

8. The volume of the gas phase is calculated by subtracting the sum of liquid volume and solid volume from the total volume.

9. Magnifying lenses are used to measure the height of phase as may be needed.

10. When there is a nonuniform distribution of phases along the pipe such as the slug flow in which long liquid slugs and large gas bubbles flow alternately, the length of the measuring section must be taken long enough to contain several units of the nonuniform distribution. The distance between two valves is an important value to obtain the precise experimental results.

11. Reproducibility of experimental data must be checked by many closing actions of the valves under one experimental condition. One instantaneous volume-averaged value of the volumetric fraction is obtained by one closing action of valves. The mean value of the volume-averaged value of the volumetric fraction can be obtained as the mean value of experimental data by many closing actions. One example of the number of closing actions and their mean value is shown in Fig. 30.4 along with the experimental value obtained by each closing action.

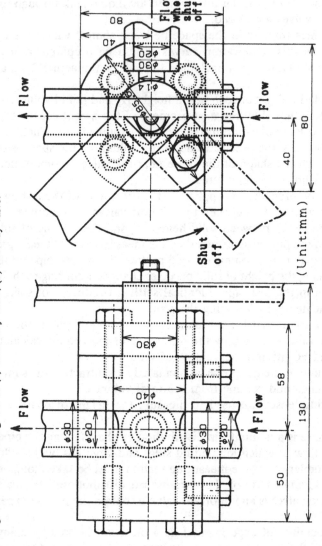

Fig. 30.2. Quick-closing valves: (a) three-way valve; (b) two-way valve.

Flow

Flow when shut off

Shut off

Flow

Flow

80

40

φ20

φ30

φ14

R85

80

40

(Unit:mm)

φ30

φ40

φ30

φ20

φ30

φ20

Flow

Flow

58

130

50

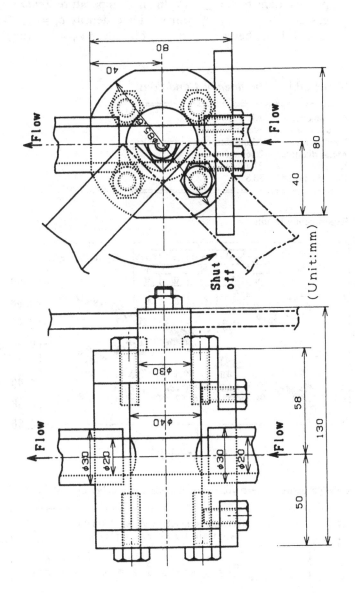

Flow

Flow

Shut off

(Unit:mm)

Flow

Flow

φ30

φ40

φ30

φ20

φ30

φ20

130

58

50

80

40

80

40

φ85

20

Suggested headings

Constants: pipe diameter $D =$ ___ (mm);   particle diameter $d_s =$ ___ (mm)
particle density $\rho_s =$ ___ (kg/m^3);   temperature *Temp* = ___ (°C)
gas density $\rho_G =$ ___ (kg/m^3);   liquid density $\rho_L =$ ___ (kg/m^3)
total volume between two shut off valves $V_T =$ ___ (m^3)

Table 30.1. *Experimental Conditions*

Gas phase:	Air
Liquid phase:	Water
Solid phase:	Aluminum ceramic particles
Mean diameter:	2.56 mm
Density:	2380 kg/m^3
Pipe diameter:	30.8 mm

Fig. 30.3. Experimental results.

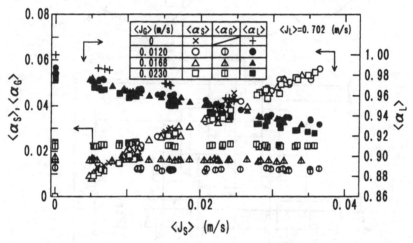

Fig. 30.4. Relation between mean volumetric fraction and number of closings of valves.

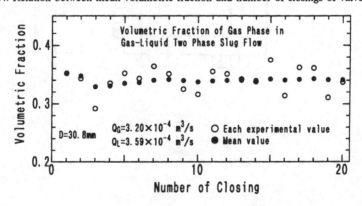

Flow conditions: $P_u =$ ___ (Pa); $Q_G =$ ___ (m³/s)

$v_L =$ ___ (m³); $v_s =$ ___ (m³); $T =$ ___ (sec)

$Q_L =$ ___ (m³/s); $Q_s =$ ___ (m³/s); $Q_T =$ ___ (m³/s)

$J_G =$ ___ (m/s); $J_L =$ ___ (m/s); $J_s =$ ___ (m/s)

$\beta_G =$ ___ ; $\beta_L =$ ___ ; $\beta_s =$ ___

Measured results:

H_L (m)	H_s (m)	H_G (m)	V_L (m³)	V_s (m³)	V_G (m³)	α_L	α_s	α_G	ρ (kg/m³)

References

1. Hetsroni, G. (ed.), *Handbook of Multiphase Systems*, McGraw-Hill, New York, 1982.
2. Hewitt, G. F., *Measurement of Two Phase Flow Parameters*, Academic, New York, 1978.

Tadashi Sakaguchi

Tadashi Sakaguchi is Professor of Multiphase Flow
Engineering at the Department of Mechanical
Engineering, Kobe University, Japan. He was a
research associate at Kobe University in 1962, an
associate professor at Kobe University in 1966, and
a visiting professor at the Swiss Federal Institute
of Technology, Zurich, 1973–4, and has been a
professor at Kobe University since 1976.

EXPERIMENT 31
Measurement of the latent heat of vaporization of a liquid

Contributed by
EFSTATHIOS E. MICHAELIDES

Principle

The use of the Clausius–Clapeyron equation is for the calculation of the latent heat of vaporization from saturation pressure and temperature data.

Object

The experiment demonstrates the relationship between saturation temperature and pressure, which is one of the fundamental relationships in the theory of phase equilibria. Furthermore, calculations with the experimental data result in an approximate expression for the latent heat of the liquid. This method also demonstrates how p, v, T data can be used together with equations of thermodynamics to yield calorimetric properties that are hard to obtain experimentally.

Background

When two phases in this case vapor and liquid are in thermodynamic equilibrium the changes in their specific Gibbs free energies are the same:

$$dg' = dg'' \tag{31.1}$$

where g is the specific Gibbs free energy, and the prime indicates liquid and the double prime vapor. Temperature and pressure are intensive variables and as such are common to both phases. In terms of the intensive-variable changes Eq. (31.1) is written:

$$s'dT + v'dP = s''dT + v''dP \tag{31.2}$$

Equation (31.2) yields the Clausius–Clapeyron equation:

$$dP/dT = (s'' - s')/(v'' - v') \tag{31.3}$$

240

The entropy difference is equal to L/T, where L is the latent heat of vaporization. Furthermore, when the fluid is far from the critical state, $v'' \gg v'$ and v'' can be approximately given by the ideal gas equation ($v'' = RT/P$). Under these conditions Eq. (31.3) is reduced to the following expression:

$$\frac{dP}{dT} = \frac{PL}{RT^2} \tag{31.4}$$

which can be integrated (using the atmospheric conditions P_0 and T_0 as upper limits) to yield the following:

$$\ln\left(\frac{P}{P_0}\right) = \frac{L}{R}\left(\frac{1}{T_0} - \frac{1}{T}\right) \tag{31.5}$$

Because of the simplifying assumptions the last equation can be considered only as approximate. However in the case of water at subatmospheric pressures the error in the calculated value of L introduced by this approximation is less than 2 percent.

Typical experimental results for water-saturation pressures and temperatures are shown in Fig. 31.1, plotted on semilog paper. The latent heat of water as calculated by linear regression from these results is 2345 kJ/kg.

Fig. 31.1. Saturation equilibrium for water.

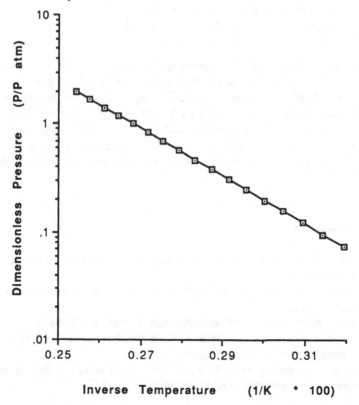

The strongest assumption in deriving Eq. (31.5) is that the latent heat is constant over the range of temperatures considered. One may relax this assumption by assuming that L is given by a polynomial function:

$$L = A + BT + CT^2 + \ldots \tag{31.6}$$

When this expression is substituted in Eq. (31.4) an integrable expression results and the logarithm of pressure is given as a polynomial of temperature. The students can obtain this polynomial and from the data points gathered they can devise a method to calculate the coefficients $A, B, C, \ldots$. There is a unique solution for these coefficients provided the number of coefficients is less than or equal to the number of pressure–temperature pairs minus one.

Apparatus

One vacuum pump
One control valve for gases
Two glass flasks (capacity ≈ 1 l), one with water or other liquid and the other
      empty
One laboratory-scale condenser
One electric or gas heater
One long mercury manometer (height ≈ 75 cm)
Connecting glass pipes and rubber vacuum seals
One thermometer and one barometer

A schematic diagram of the apparatus is shown in Fig. 31.2. The arrangement used consists of a small vacuum pump, a 60-cm-long cylindrical condenser with helical vapor passage cooled by tap water in a counterflow mode, two laboratory flasks of 1 l capacity each, and an electric heater. The control valve is used to regulate the pressure in the system by admitting more or less atmospheric air, the manometer to measure the vapor pressure in the heated flask, and a copper-constantan thermocouple to measure the temperature of the water.

Procedure

The flask is filled with distilled water and the water lines of the condenser are supplied with tap water.

The heater is switched on and with the vacuum pump shut the water is at atmospheric pressure.

The water is heated up and subsequently it boils at about 100 °C.

The absolute pressure is measured with the barometer and is recorded together with the boiling temperature.

The vacuum pump is switched on and the control valve is adjusted to maintain constant subatmospheric pressure in the system.

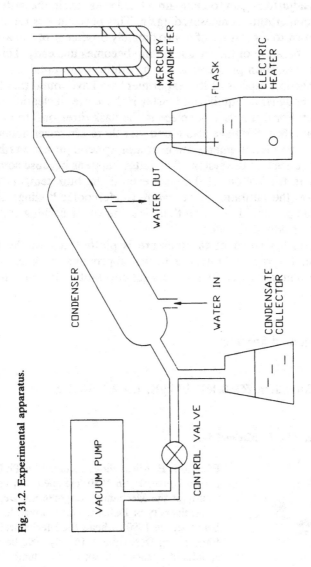

Fig. 31.2. Experimental apparatus.

MERCURY MANOMETER

FLASK

ELECTRIC HEATER

CONDENSER

WATER OUT

WATER IN

CONDENSATE COLLECTOR

VACUUM PUMP

CONTROL VALVE

The mercury manometer indicates the difference between the atmospheric pressure and the pressure in the system.

The water now boils at a lower temperature which is recorded together with the system's pressure (saturation temperature and pressure). The control valve is adjusted again to maintain a third pressure in the system and the saturation temperature is measured again. This procedure is repeated about ten times down to a pressure of approximately 5 percent of the atmosphere, or until the operation of the vacuum pump becomes unsteady. Thus, close to ten pairs of saturation pressures and temperatures are obtained.

On the practical aspects of this experiment we have found that the system attains thermodynamic equilibrium faster if there are nucleation sites in the boiling water. For this reason we place in the flask three or four bolts or nuts or some chalk dust. We have also found that there are fewer nonequilibrium effects if the first measurements are near atmospheric pressure rather than at the lowest pressure of the system. The latter happens because some bubbles may appear in the bottom of the flask even if the bulk temperature of the water is below the saturation temperature (undercooled boiling). However if the saturation pressure is lowered there is some vapor flashing and the water temperature is adjusted faster.

The natural logarithm of the pressure is plotted against the inverse of temperature. The curve obtained is to first approximation linear. The slope of the line when multiplied by the ideal gas constant yields the latent heat of vaporization.

Suggested headings

Constants: for water, $R = 0.4619$ kJ/kgK, $L = 2,345$ kJ/kg

Efstathios E. Michaelides

Efstathios E. Michaelides received his Ph.D from Brown University in 1980. He spent ten years in the faculty of the Mechanical Engineering Department at the University of Delaware and moved to Tulane University in 1990 to head the Mechanical Engineering Department. In July 1992 he was appointed Associate Dean for Graduate Studies and Research. He teaches courses in the thermal sciences and conducts research in the areas of particulate flows, multiphase flows, turbulence modification by a dispersed phase, and geothermal energy conversion. He is the author of more than 100 scientific research papers in journals and conference proceedings.

EXPERIMENT 32

Dilution techniques for the performance evaluation of continuous-flow combustion systems

Contributed by
ARTHUR H. LEFEBVRE

Object

In the design of continuous-flow combustion systems, an important performance requirement is that combustion must be sustained over a wide range of
operating conditions. For the combustors employed in aircraft gas turbines,
this poses special problems because they are often called upon to operate at
very low inlet temperatures and pressures and at fuel/air ratios that lie well
outside the normal burning limits of hydrocarbon–air mixtures.

The stability performance of a continuous-flow combustor is usually expressed in the form of a stability plot that separates the regions of stable and
unstable combustion. The traditional plot has equivalence ratio or fuel/air
ratio as the ordinate, and some loading parameter, such as air velocity or air
mass flow rate through the combustor, as the abscissa. A plot of this type is
often called a stability *loop*, owing to its shape, as illustrated in Fig. 32.1.

Background

Stability loops provide two basic kinds of information. First, for any given
fuel/air ratio, they indicate the blowout velocity U_{BO}, which is the gas velocity
at which flame extinction occurs. Attention is usually focused on the maximum blowout velocity, which tends to coincide with mixture strengths that lie
close to the stoichiometric value. Second, for any given combustor loading,
they show the range of fuel/air ratios over which stable combustion can be
achieved.

A widely used method for stabilizing a flame in a flowing stream of com-
bustible mixture is by the insertion of a bluff object such as a disk, cone, or
"Vee" gutter, which produces in its wake a low-velocity recirculatory flow
region in which combustion can be initiated and sustained. Due to apparatus
limitations, in particular the difficulty and high cost of providing high air flow
rates at low (subatmospheric) pressures, most of the reported experimental
studies on bluff-body flame stabilization have used air supplied at normal

245

atmospheric pressure. Thus, when the flame holders tested have been of a practical size, the results have usually been confined to very weak or very rich fuel/air mixtures. This point is well illustrated by the stability data plotted in Fig. 32.2. When tests have been carried out in the most interesting range of fuel/air ratios, that is, near stoichiometric, either velocities have been very high or dimensions have been very small. Extrapolation of the experimental data to practical velocities or practical dimensions is a somewhat dubious process. It is difficult to extrapolate dimensions, because any such extrapolation must also take into account effects arising from a change in "blockage." It is equally difficult to extrapolate velocities, because at high velocities compressibility effects can change the flow pattern in and around the combustion zone.

The water injection technique has the enormous advantage of allowing full-scale combustors to be fully evaluated at low cost while operating within their normal range of velocities and fuel/air ratios. Fan air is used, and low pressures are simulated by introducing water into the combustion zone. The essence of the method is the theoretical equivalence, on a global-reaction-rate basis, between a reduction in reaction pressure and a reduction in reaction temperature (which in this instance is accomplished by the addition of water).

Fig. 32.1. Typical combustion-chamber stability loop.

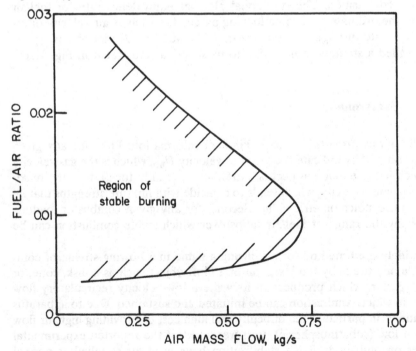

Apparatus

One of the most useful applications of this technique is in obtaining blowout
data for various designs of flame holder. The apparatus employed is shown
schematically in Fig. 32.3. Essentially, it comprises
a supply of air at atmospheric pressure,
a preheat combustion chamber,
a working section containing the flame holder under test,
and provisions for injecting liquid fuel and water in well-atomized form into
the flowing gas upstream of the flame holder.
   Sufficient time and temperature are provided between the planes of injection

Fig. 32.2. Effect of mixture velocity and baffle diameter on stability range $P_a = 1$ atmosphere.[3]

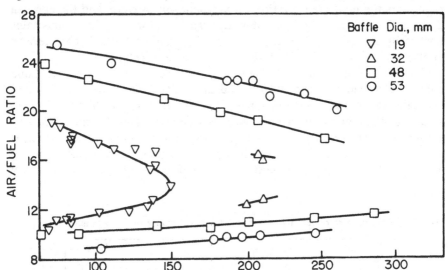

Fig. 32.3. Schematic diagram of apparatus requirements for water-injection technique.

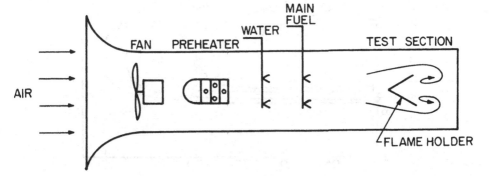

of water and fuel and the flame holder to ensure that both liquids are fully prevaporized and premixed upstream of the reaction zone.

Procedure

The test procedure is quite simple.

The velocity and temperature of the gas flowing over the stabilizer are adjusted to the desired values, the fuel is turned on, and an ignition source is used to ignite the flame in the wake region downstream of the stabilizer.

Once the flame is established, the water is turned on and the flow of water is gradually increased until the flame goes out. This process is repeated at a sufficient number of fuel flow rates for a complete stability loop to be drawn. Some typical stability loops are shown in Fig. 32.4, in which the ordinate represents the equivalence ratio of a kerosene–air mixture, and the abscissa denotes the mass ratio of water flow to kerosene flow. This figure illustrates how easily and clearly the point of peak stability, a most useful characteristic of stability loops, is defined by the water-injection technique.

Figure 32.4 was obtained by Rao and Lefebvre[5] using a Vee-gutter flame holder 6.2-cm wide and with 30° included angle. The flame holder was mounted vertically at the center of a rectangular test section, of size 0.15 m × 0.2 m, with its apex pointing upstream. Figure 32.4 demonstrates that an increase in approach stream velocity has an adverse effect on stability because it reduces the residence time of the reactants in the wake region. The influence of flame holder forebody shape on stability is illustrated in Fig. 32.5, in which stability

Fig. 32.4. Effect of mixture velocity on the stability performance of 30° Vee gutter.[5]

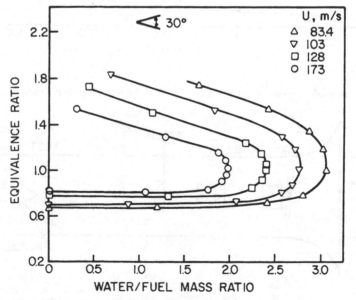

data are plotted for five different flame holder configurations. The five Vee gutters represented in this figure are all 4 cm in width and have included angles of 30, 45, 60, 90, and 180 degrees. The beneficial effect on stability of an increase in included angle is very apparent in this figure.

The water-injection technique also lends itself to the study of ignition phenomena in continuous-flow combustion systems. Lefebvre and Halls[2] used this method to determine the relative merits of various designs of surface-discharge igniters. The test section employed is shown in Fig. 32.6. It comprises a length of 10-cm-diameter piping with the igniter located about 30 cm from its open end. The igniter face is arranged to be flush with the inside surface of a short cylindrical tube that is mounted concentrically within the 10-cm pipe and is incorporated in order to present the plug face to midstream gas where composition and velocity are near to the measured mean values (i.e., boundary-layer effects are avoided). Initial preheating of the air to 500 °C ensures that the fuel and water are completely vaporized in the plane of the igniter. The test procedure is similar to that employed in measuring stability. Water is gradually admixed with the fuel until passage of the spark no longer produces a streak of flame downstream of the plug.

Results

Ignition loops for three different designs of igniter are shown in Fig. 32.7. Again results are plotted as graphs of equivalence ratio versus water/fuel

Fig. 32.5. Effect of flame holder forebody shape on stability.[5]

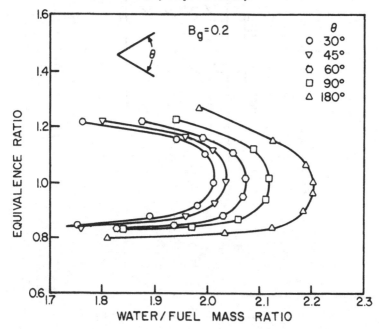

Fig. 32.6. Apparatus for determining ignition loops.[2]

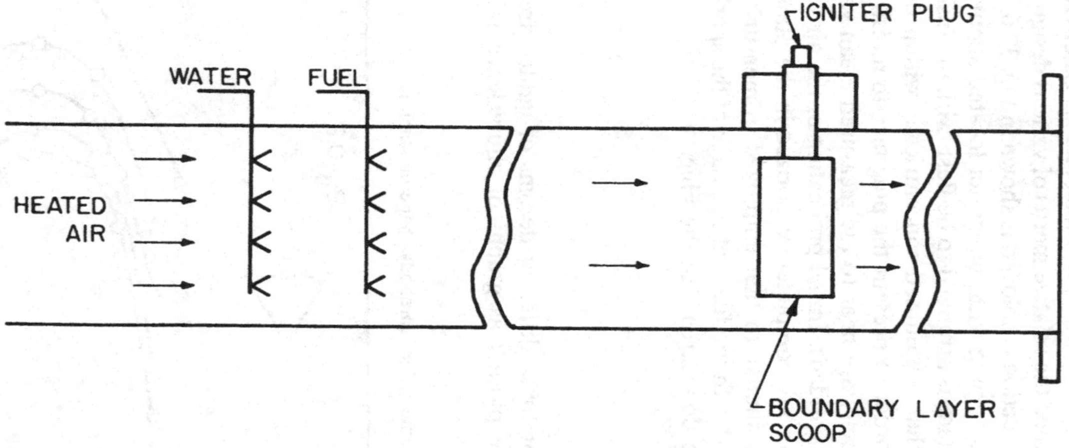

ratio. In the same figure, for comparison, are plotted ignition loops for the same three plugs when operating in an actual combustion chamber at a constant subatmospheric pressure. These loops are plotted in the conventional manner as graphs of overall equivalence ratio versus mass flow. Comparison of the two sets of curves shows good qualitative agreement.

A most useful asset of the water-injection technique is that it enables various important operating parameters, such as inlet air temperature and velocity, to be examined for their effects on ignition and stability performance while using practical high-energy ignition devices and large-scale flame holders, respectively.

Curves of the type shown in Fig. 32.5 also provide useful data whereby the

Fig. 32.7. Ignition loops for three different designs of igniter plug.[2]

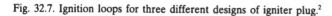
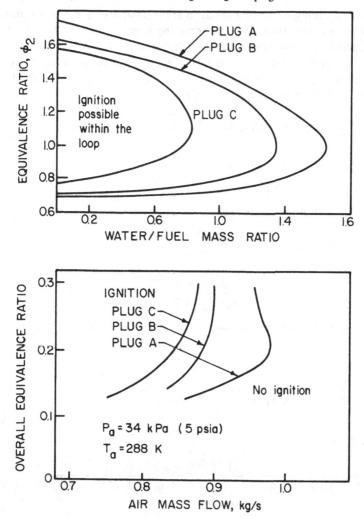

basic stability of various designs of gutter may be compared. The only assumption involved is a reasonable one, namely, that the gutter requiring the largest amount of water to cause flame extinction has the best stability. The value of the technique is further enhanced by a relationship (derived from global-reaction-rate considerations) between the fraction of water in the fuel and the equivalent reduction in gas pressure. This aspect is especially useful for studying the performance of aircraft combustion systems which are sometimes called upon to operate at pressures as low as 21 kPa (3 psia).

The air loading on the combustion zone can be expressed in terms of the reaction temperature and the concentrations of fuel and oxygen by a global-reaction-rate equation of the form

$$\frac{\dot{m}_a}{VP^n} \propto \exp\{-(E/RT)\} \frac{\phi^{m-1}}{\beta T^{n-0.5}} x_f^m x_o^{n-m} \tag{32.1}$$

The effect of adding water to a reaction zone is twofold. First, by its presence, it reduces the concentrations of the reacting species. Second, by virtue of its heat capacity, it lowers the reaction temperature. By substituting into Eq. (32.1) appropriate expressions for x_f and x_o in terms of the molar concentrations of all the species present in the reaction zone, we obtain for the loading function for fuel-weak mixtures ($\phi_2 < 1$):

$$\frac{\dot{m}_a}{VP^2} \propto \exp\{-(E/RT)\} \frac{(\phi_2 - \phi_1)^{m-1}}{\beta T^{n-0.5}} \times$$

$$\frac{(1-\beta)^m [1 - \phi_1 - \beta(\phi_2 - \phi_1)]^{n-m}}{[85.7 + 6\phi_2 - 5(1 - 1.87K - \beta)(\phi_2 - \phi_1)]^n} \tag{32.2}$$

The corresponding equation for rich mixtures ($\phi_2 > 1$) is

$$\frac{\dot{m}_a}{VP^n} \propto \exp\{-(E/RT)\} \frac{(\phi_2 - \phi_1)^{n-1}}{\beta T^{n-0.5}} \times$$

$$\frac{(1-\beta)^n (1 - \phi_1)^{n-m}}{[85.7 + 6\phi_1 - 18\beta(1 - \phi_1) + (1 + 23\beta + 9.345K)(\phi_2 - \phi_1)]^n} \tag{32.3}$$

For a bimolecular reaction in which $m = 1$ and $n = 2$, Eqs. (32.2) and (32.3) become respectively, for $\phi_2 < 1$,

$$\frac{\dot{m}_a}{VP^2} \propto \frac{(1-\beta) [1-\phi_1-\beta(\phi_2 - \phi_1)]}{T^{1.5} \exp(E/RT)\beta[(85.7+6\phi_2)-5(1-1.87K-\beta)(\phi_2-\phi_1)]^2} \tag{32.4}$$

and, for $\phi_2 > 1$,

$$\frac{\dot{m}_a}{VP^2} \propto \frac{(1-\beta)^2(\phi_2-\phi_1)(1-\phi_1)}{T^{1.5} \exp(E/RT)\beta[(85.7+6\phi_1-18\beta(1-\phi_1)+(1+23\beta+9.34K)(\phi_2-\phi_1)]^2} \tag{32.5}$$

The previous equations describe the performance of a homogeneous reactor supplied with fresh or vitiated air plus kerosene fuel of which a fraction $(1 - \beta)$ leaves the reactor unburned. ϕ_1 is the equivalence ratio of combustion processes occurring upstream of the reactor; it is assumed that any such process takes place with 100 percent combustion efficiency. ϕ_2 is the overall equivalence ratio. The reacting species are unburned fuel and oxygen, and the effect of water in reducing their concentration is taken into account in the Eqs. (32.3)–(32.5) where the quantity of added water is expressed as a fraction K of the fuel supplied to the reactor.

The reaction temperature T is obtained by a heat balance. The heat lost by the completely burned products in falling from the adiabatic flame temperature to the reaction temperature is equated to the heat absorbed in raising the unburned fuel and air to the reaction temperature and in converting water or steam at inlet temperature to steam at the reaction temperature. Full allowance is made for the effects of chemical dissociation.

Upon substitution of appropriate values of ϕ_1, ϕ_2, and β into the previous equations, together with the corresponding calculated values of T, relationships may be obtained between the quantity of added water and the equivalent fall in gas pressure. Such a relationship is shown in Fig. 32.8 for a value of E of 42,000 calories per mole and a preheater equivalence ratio ϕ_1 of 0.2. It may be observed in this figure that the effect of adding a pound of water for each pound of kerosene is roughly equivalent to halving the combustion pressure.

The use of water as a diluent in the combustion process has been criticized on the grounds that water does not remain inert at the high temperatures produced by the combustion of stoichiometric mixtures. Fortunately, the dissociation of water into hydrogen and oxygen, and the resulting production of

Fig. 32.8. Relationship between water dilution and effective decrease in pressure.[6]

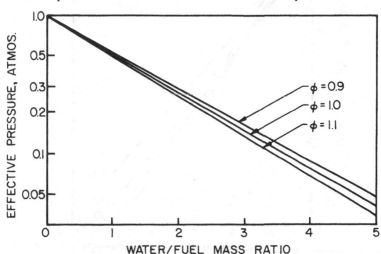

radicals which affect chemical reaction rates, does not occur to any great extent because the addition of water reduces the reaction temperature to a level where dissociation is slight.

The main drawback to the water-injection technique is that a preheat combustion chamber is needed to vaporize the injected water and the main fuel. Usually it is necessary to raise the fan outlet temperature from around 300 K to around 700 to 850 K to ensure that the water and fuel are completely vaporized upstream of the test section. Because of this drawback to the use of water, alternative diluents have occasionally been considered.[1,4,6] One obvious candidate, which remains inert to much higher temperatures than water, is nitrogen. The main advantage of nitrogen over water is that no preheating is required. Its main drawback is that it is expensive and, due to its much lower heat capacity, the mass-flow requirements are much larger than for water. If large quantities of nitrogen are contemplated, a liquid storage system would reduce the nitrogen cost considerably, but would require special installation and handling equipment. The cost of nitrogen is, of course, relative to the scale of combustion device on which the technique is employed. Hence, for small-scale laboratory work, the use of nitrogen can sometimes be more economical overall than water.

Another drawback to the nitrogen-dilution technique is that it is limited to gaseous fuels. Norster[4] has calculated the relationship between the quantity of added nitrogen and the equivalent reduction in pressure for propane–air mixtures. His results are shown in Fig. 32.9. They show that from 3.0 to 3.5

Fig. 32.9. Relationship between nitrogen dilution and effective decrease in pressure.[4]

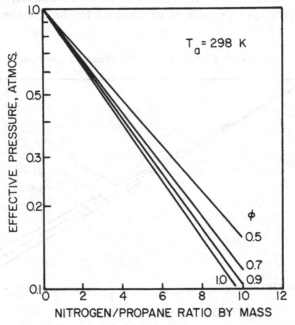

pounds of nitrogen are required for each pound of propane in order to effectively halve the combustion pressure.

In conclusion, the following results are significant:

1. The water-injection technique constitutes a cheap and convenient method of closing stability and ignition loops, and is generally useful for assessing and comparing the performance of full-scale combustion systems over a wide range of conditions.
2. A reduction in combustion pressure to any desired subatmospheric value is readily simulated by varying the quantity of injected water.
3. The technique is unsuitable for application to situations where combustion performance is limited by mixing or fuel atomization or any process other than chemical reaction rates.
4. The use of gaseous nitrogen instead of water eliminates the need for a preheat combustion chamber. The main drawbacks to the use of nitrogen are that it is considerably more expensive than water and is restricted to gaseous fuels. Its application is limited therefore to small-scale laboratory experiments.

References

1. Lefebvre, A. H., "Some simple techniques for the performance evaluation of gas turbine combustion systems," *Experimental Methods in Combustion Research,* ed. J. Surugue, pp. 5–21, Pergamon, New York, 1961.
2. Lefebvre, A. H., and Halls, G. A., "Simulation of low combustion pressures by water injection," *Seventh Symposium (International) on Combustion,* pp. 654–8 The Combustion Institute, Pittsburgh, 1958.
3. Longwell, J. P.; Chenevey, J. E.; Clark, W. W., and Frost, E. E., "Flame stabilization by baffles in a high velocity gas stream," *Third Symposium (International) on Combustion,* pp. 40–4, Williams and Wilkins, 1949.
4. Norster, E. R., "Subsonic flow flameholder studies using a low pressure simulation technique," *Combustion in Advanced Gas Turbine Systems,* ed. I. E. Smith, Cranfield International Symposium Series, vol. 10, pp. 79–91, Pergamon, New York, 1967.
5. Rao, K. V. L., and Lefebvre, A. H., "Flame blowoff studies using large-scale flameholders," *ASME Journal of Engineering for Power* 104 (1982): 853–7.
6. Stwalley, R. M., "Flame stabilization by bodies of irregular shape," M.S. Thesis, School of Mechanical Engineering, Purdue University, 1985.

Notation

E	activation energy, cal./mole
K	water/fuel mass ratio
$\dot{m}_a$	air mass flow rate, kg/s
m	exponent of fuel concentration
n	reaction order
P	pressure, Pa
R	gas constant (1.986 cal./mole · K)
T	reaction temperature
U	mixture velocity, m/s

V combustion volume, m^3
x_f molar fraction of fuel
x_o molar fraction of oxygen
β fraction of fuel burned
ϕ_1 equivalence ratio of preheat combustion process
ϕ_2 overall equivalence ratio

Arthur Lefebvre

Arthur Lefebvre was educated at Nottingham University and Imperial College, London, where he obtained Ph.D. and D.Sc. degrees in mechanical engineering. His industrial experience includes ten years with the Aero-Engine Division of Rolls Royce at Derby, where he worked on combustion research and the design and performance analysis of gas-turbine combustion systems. He has served as Head of Mechanical Engineering at the Cranfield Institute of Technology in England and is now Reilly Professor of Combustion Engineering at Purdue University. Dr. Lefebvre holds a large number of patents relating to atomizers and combustion equipment. His publications include papers and books on both fundamental and practical aspects of atomization and combustion.

APPENDIX 1
Experiments and demonstrations in thermodynamics

The following list of experiments and demonstrations is presented to supplement those given in the body of this book. Though they use different formats than that used herein, there is sufficient information and challenge for students to easily assemble the apparatus and conduct measurements that will assist them in understanding more about thermodynamics.

1.A Properties*

1.A.1   "Determination of the specific heat ratio of air by standing waves," by H. W. Butler, Ref. 3, p. 201.
1.A.2   "Critical point demonstration," by P. J. Waibler, Ref. 4, p. 206.
1.A.3   "Demonstration of the anisotropic thermal conductivity of wood," by S. Corrsin, Ref. 4, p. 229.
1.A.4   "Errors in temperature measurements," by E. A. Brun, Ref. 2, p. 109.
1.A.5   "The meaning of temperature," by R. Baierlein, *Phys. T.* 28 (1990): 94.
1.A.6   "The thermal properties of materials," by J. Ziman, *Sci. Am.* 217, 3 (1967): 181.
1.A.7   "How to measure vapor pressure," by C. L. Strong, *Sci. Am.* 223, 6 (1970): 116.
1.A.8   "How salt intensifies cooling," by J. Walker, *Sci. Am.* 250, 4 (1984): 150.
1.A.9   "Measurement of a thermodynamic constant (k)," by W. Connolly, *Phys. T.* 26 (1988): 235.
1.A.10 "Demonstration of the thermal expansion of solids," by S. K. Chakarvarti, *Phys. T.* 26 (1988): 400.
1.A.11 "Boiling water and the height of mountains," by J. P. Negret, *Phys. T.* 24 (1986): 290.

*Note: *I.J.M.E.E.* = *The International Journal of Mechanical Engineering Education*;
  Phys. T. = *Physics Teacher*;
  Sci. Am. = *Scientific American*.

257

1.A.12   "Heat of vaporization of water," by P. O. Berge and J. E. Huebler, *Phys. T.* 16 (1978): 476.

1.A.13   Measuring the partial pressures of water vapor," by M. K. Smith, *Phys. T.* 16 (1978): 476.

1.A.14   "A dewar for the heat of vaporization of liquid nitrogen," by P. O. Berge, J. R. Shipman, and J. E. Huebler, *Phys. T.* 14 (1976): 571. (See also 1.A.18.)

1.A.15   "Demonstration of a crystalline phase change in a solid," by A. A. Bartlett, *Phys. T.* 13 (1975): 545.

1.A.16   "Is water the only substance which expands (at some temperature) when cooled?," by W. Thumm, *Phys. T.* 13 (1975): 290.

1.A.17   "Projection of freezing by boiling phenomena (triple point)," by J. A. Davis, *Phys. T.* 13 (1975): 180.

1.A.18   "A simple measurement of the heat of vaporization of liquid nitrogen," by P. A. Knutsen and G. A. Salinger, *Phys. T.* 7 (1969): 288.

1.A.19   "Latent heat," by J. Harrin and A. Ahlgren, *Phys. T.* 4 (1966): 317.

1.A.20   "Sublimation of ice," by F. W. Kantov, *Phys. T.* 3 (1965): 322.

1.A.21   "Measurement of the coefficient of expansion of a liquid," by D. S. Ainslie, *Phys. T.* 2 (1964): 338.

1.A.22   "Instructional experiment for compressibility and vapour pressure," by R. A. Gaggioli and W. J. Wepfer, *I.J.M.E.E.* 9, 2 (1981): 111.

1.B   Instrumentation and apparatus

1.B.1   "Thermocouple lead conduction errors," by F. Landis, Ref. 4, p. 108.

1.B.2   "Ruchhardt's apparatus for determination of C_p/C_v of a Gas," by D. W. Batteau and B. Watson, Ref. 4, p. 203.

1.B.3   "Shielding of thermocouples," by A. L. London, Ref. 4, p. 243.

1.B.4   "Calibration and efficiency of microwave ovens," by R. Fritz, *Phys. T.* 28 (1990): 564.

1.B.5   "How to make a temperature control," by C. L. Strong, *Sci. Am.* 213, 4 (1965): 106.

1.B.6   "A simple specific heat apparatus," by T. W. Geisert, *Phys. T.* 21 (1983): 619.

1.B.7   "A simple temperature sensor," by S. K. Chakarvarti, *Phys. T.* 20 (1982): 470.

1.B.8   "Thermocouple temperature unit," by C. Goddard and M. Pavalow, *Phys. T.* 13 (1975): 437.

1.B.9   "Inexpensive immersion heater," by F. G. Kariorsis, *Phys. T.* 12 (1974): 573.

1.C   Thermometers

1.C.1   "Pneumatic gas thermometer," by H. W. Butler, Ref. 4, p. 110.

1.C.2   "Thermometer calibration and temperature measurement," by E. A. Brun, Ref. 2, p. 107.

1.C.3 "Time constant of a thermometer," by V. Zanetti, *Phys. T.* 21 (1983): 108.

1.C.4 "Thermometer from a Bic ballpoint pen," by A. R. Iversin, *Phys. T.* 15 (1977): 186.

1.C.5 "Temperature and the transfer of heat," by C. T. Haywood, *Phys. T.* 14 (1976): 366.

1.D Gases

1.D.1 "The physics of popping popcorn," by R. G. Hunt, *Phys. T.* 29 (1991): 230.

1.D.2 "Brownian Motion," by B. H. Lavenda, *Sci. Am.* 252, 2 (1985): 70.

1.D.3 "Trace gases, CO_2, climate and the greenhouse effect," by G. Aubrecht, *Phys. T.* 26 (1988): 145.

1.D.4 "Visualization of the ideal gas equation," by J. Hellemans, *Phys. T.* 26 (1988): 398.

1.D.5 "Boyle's law projected," by E. Zwicker and L. Alofs, *Phys. T.* 24 (1986): 118. (See 1.D.7.)

1.D.6 "Gas-law apparatus made of copper water pipe," by G. Brown, *Phys. T.* 12 (1974): 306.

1.D.7 "Boyle's law demonstration using a vacuum gauge," by W. Carlson, *Phys. T.* 5 (1967): 387.

1.E Entropy

1.E.1 "Determination of Boltzmann's constant," by E. V. Lee, *Phys. T.* 13 (1975): 305.

1.E.2 "Maxwell's demon," by J. R. Larson, *Phys. T.* 13 (1975): 503.

1.F 1st law

1.F.1 "Energy and power," by C. Starr, *Sci. Am.* 224, 3 (1971): 37.

1.F.2 "The conversion of energy," by C. M. Summers, *Sci. Am.* 224, 3 (1971): 148.

1.F.3 "Energy and the automobile," by G. Waring, *Phys. T.* 18 (1980): 494.

1.F.4 "Geothermal energy: the furnace in the basement," by M. C. Smith, *Phys. T.* 16 (1978): 533.

1.F.5 "An instructional experiment for first and second law analysis of a gas-fired heater," by R. A. Gaggioli and W. J. Wepfer, *I.J.M.E.E.* 9, 4 (1981): 283.

1.G Cycles and engines

1.G.1 "The Stirling engine – 173 years old and running," by H. R. Crane, *Phys. T.* 28 (1990): 252.

1.G.2 "The heat pump," by J. F. Sandfort, *Sci. Am.* 184, 5 (1951): 54.

1.G.3   "Rudolf Diesel and his rational engine," by L. Bryant, *Sci. Am.* 221, 2 (1969): 108.

1.G.4   "The origins of the steam engine," by E. S. Ferguson, *Sci. Am.* 210, 1 (1964): 98.

1.G.5   "The Philips air engine," by L. Engel, *Sci. Am.* 179, 1 (1948): 52.

1.G.6   "Rotary engines," by W. Chinitz, *Sci. Am.* 220, 2 (1969): 90.

1.G.7   "Construction of lighter-flint heat engine," by C. L. Strong, *Sci. Am.* 201, 1 (1959): 145.

1.G.8   "Construction of a rubber-band heat engine," by C. L. Strong, *Sci. Am.* 224, 4 (1971): 118.

1.G.9   "How to make refrigeration machines," by C. L. Strong, *Sci. Am.* 221, 5 (1969): 151.

1.G.10   "How to construct a simple thermal engine," C. L. Strong, *Sci. Am.* 225, 1 (1971): 114.

1.G.11   "The Parsons steam turbine," by W. G. Scaife, *Sci. Am.* 252, 4 (1985): 132.

1.G.12   "Experiments with the external-combustion fluidyne engine which has liquid pistons," by J. Walker, *Sci. Am.* 252, 4 (1985): 140.

1.G.13   "Rotary Curie-point heat engine," by G. Barnes, *Phys. T.* 24 (1986): 204.

1.G.14   "Applications of refrigeration systems," by A. Bartlett, *Phys. T.* 24 (1986): 92.

1.G.15   "Steam engine efficiency," by R. Euclids and S. Welty, *Phys. T.* 24 (1986): 308.

1.G.16   "Power output of the Otto cycle engine," by E. Zebrowski, *Phys. T.* 22 (1984): 390.

1.G.17   "Thermal expansion heat engine," by C. H. Blanchard, *Phys. T.* 21 (1983): 319.

1.G.18   "Real Otto and diesel engine cycles," by R. Giedd, *Phys. T.* 21 (1983): 29.

1.G.19   "Stirling engines for demonstrations," by R. D. Spencer and C. L. Foiles, *Phys. T.* 20 (1982): 38.

1.G.20   "A model of the Savery steam engine," by G. W. Ficken, *Phys. T.* 19 (1981): 228.

1.G.21   "Solid state solar engine," by J. Jedlicka, *Phys. T.* 10 (1972): 475.

1.G.22   "An experimental vapour compression demonstration unit for studies in refrigeration and air conditioning," by J. R. Ghidella, M. K. Raff, and K. Srinivasan, *I.J.M.E.E.* 19, 3 (1991): 229.

1.H   Low-temperature thermodynamics

1.H.1   "Two cryogenic demonstrations," by T. K. McCarthy, *Phys. T.* 29 (1991): 575.

1.H.2   "Determining absolute zero in the kitchen," by R. Otani and P. Siegel, *Phys. T.* 29 (1991): 316.

1.H.3 "Low temperature physics," by H. M. Davis, *Sci. Am.* 180, 6 (1949): 30.
1.H.4 "How to liquify gases," by C. L. Strong, *Sci. Am.* 221, 5 (1969): 151.
1.H.5 "The spectroscopy of supercooled gases," by D. H. Levy, *Sci. Am.* 250, 2 (1984): 96.

1.I Miscellaneous

1.I.1 "Hot water freezes faster than cold water," by J. Walker, *Sci. Am.* 237, 3 (1977): 246–57.
1.I.2 "Demonstrating adiabatic temperature changes," by R. D. Russell, *Phys. T.* 25 (1987): 450.
1.I.3 "Atoms and molecules in small aggregate – the fifth state of matter," by G. Stein, *Phys. T.* 17 (1979): 503.
1.I.4 "Freezing by boiling using a low capacity pump without acid," by M. Graham, *Phys. T.* 15 (1977): 367.
1.I.5 "Research at very high pressures and high temperatures," by F. P. Bundy, *Phys. T.* 15 (1977): 461.
1.I.6 "Ice cube regelation," by W. A. Hilton, *Phys. T.* 12 (1974): 308.
1.I.7 "Regelation of ice is a complicated problem," by M. Zemansky, *Phys. T.* 3 (1965): 301.
1.I.8 "A modern introduction to classical thermodynamics," by E. F. Lype, *I.J.M.E.E.* 8, 4 (1980): 181.
1.I.9 "The development of an experimental facility for application at undergraduate level in the area of thermo-fluids," by A. Evans, J. Hall, and A. Henderson, *I.J.M.E.E.* 11, 4 (1983): 233.

1.J Source texts of demonstrations and experiments

Source	Topic	# Expts.	# Demos.
Ref. 1	Thermal expansion	4	
	Temperature measurement	7	
	Kinetic theory of gases	5	
	Temperature control	2	
	Low-temperature physics	3	
Ref. 5	Temperature change		8
	Specific heats		1
	Heats of transformation (superheat, etc.)		18
	Second law		2
	Gases		27
	Kinetic theory		15
	Liquification of a gas		3
	Low-temperature physics		7

In addition Sutton, Ref. 7, has numerous demonstrations on thermometry, expansion, specific heats, change of state, vapor pressure, and low-temperature physics. These demonstrations are at a low undergraduate level but serve nicely to introduce various topics in thermodynamics.

Taffel et al., Ref. 8, do not have the variety of Sutton, but do treat nice laboratory exercises on such topics as Boyle's law, the estimation of absolute zero, and the heat of vaporization of water.

Walker, Ref. 9, has collected approximately 200 references and 116 demonstrations on thermodynamics and heat transfer. The demonstrations are practical and fun, requiring less than two minutes to perform. Many of these demonstrations can spruce up an otherwise dull lecture, as they focus on familiar phenomena in an unconventional but entertaining manner.

The subject of experimentation cannot end without some reference regarding how students should present their data. Two texts are highly recommended. Granger, Ref. 3, presents a number of significant points the student should consider in writing a proper engineering report. Appendix C of Ref. 3 discusses in great detail the uncertainty in measurements and the correct procedure in calculating errors in calibration, data calibration, data acquisitions, and data reduction. Examples are presented illustrating bias, precision, and degrees of freedom.

Squires, Ref. 6, is recommended for students' study if they plan on writing a technical paper for a scientific refereed journal.

References

1. Brown, T. R. (ed.), *The Lloyd William Taylor Manual of Advanced Undergraduate Experiments in Physics*, Addison-Wesley, Reading, MA, 1959.
2. Brun, E. A., *Modern Research Laboratories for Heat and Mass Transfer*, Unesco Press, Paris, 1975.
3. Granger, R. A., *Experiments in Fluid Mechanics*, Holt, Rinehart, and Winston, New York, 1988.
4. Landis, F., *Laboratory Experiments and Demonstrations in Fluid Mechanics and Heat Transfer*, Dept. of Mech. Engr., New York Univ., New York, 1964.
5. Meiners, H. E. (ed.), *Physics Demonstration Experiments*, Ronald Press, New York, 1970.
6. Squires, G. L., *Practical Physics*, Cambridge Univ. Press, Cambridge, 1985.
7. Sutton, R. M., *Demonstration Experiments in Physics,* McGraw-Hill, New York, 1938.
8. Taffel, A.; Baumel, A., and Landecker, L., *Laboratory Manual Physics*, Allyn and Bacon, Boston, MA, 1966.
9. Walker, J., *The Flying Circus of Physics*, Wiley, New York, 1975.
10. Whittle, R. M., and Yarwood, J., *Experimental Physics for Students*, Chapman and Hall, London, 1973.

APPENDIX 2
Experiments and demonstrations in heat transfer

The following list of experiments and demonstrations is presented to supplement those given in Part I of the book. The experiments have varying degrees of difficulty and should offer additional variety to illustrate the fundamentals of heat transfer.

2.A Conduction

2.A.1 "Conduction of heat," by P. B. Allen, *Phys. T.* 21 (1983): 582.

2.A.2 "Insulation and rate of heat transfer," by V. D. Pynadath, *Phys. T.* 16 (1978): 379.

2.A.3 "The bimetallic strip – a quantitative experiment," by P. W. Hewson, *Phys. T.* 13 (1975): 350.

2.A.4 "Heat conduction through liquids," by J. P. Walsh, *Phys. T.* 8 (1970): 265.

2.A.5 "Thermal transients in distributed parameter systems," by P. W. McFadden, Ref. 4, p. 123.

2.A.6 "Determination of thermal diffusivities by Ängstrom's method," by R. Eichhorn, Ref. 4, p. 126.

2.A.7 "Fin demonstration," by W. C. Reynolds, Ref. 4, p. 130.

2.A.8 "Effect of material on heating fin performance," by W. C. Reynolds, Ref. 4, p. 130.

2.A.9 "Analysis of a lap-joint under longitudinal conduction," by F. Landis, Ref. 4, p. 131.

2.A.10 "Conducting paper analog for steady state heat transfer," by F. Landis, Ref. 4, p. 133.

2.A.11 "Solution of Laplace equation by analog techniques," by R. Eichhorn, Ref. 4, p. 138.

2.A.12 "Solutions of transient problems by electronic analog computers," by F. Landis, Ref. 4, p. 144.

2.A.13 "Hydraulic analog for sinusoidal transient inputs," by J. T. Anderson, Ref. 4, p. 149.

2.A.14 "Critical thickness of insulation," by J. E. Sunderland, Ref. 4, p. 233.

2.A.15   "Steady state technique for thermal conductivity measurement," by E. A. Brun, Ref. 2, p. 110.

2.A.16   "Fin heat transfer," by E. A. Brun, Ref. 2, p. 112.

2.A.17   "The conduction of heat in solids," by R. L. Sproull, *Sci. Am.* 207, 6 (1962): 92.

2.B   Convection

2.B.1   "An experimental study of forced convection over finned cylinders," by R. D. Flack, *I.J.M.E.E.* 8, 1 (1980): 43.

2.B.2   "An experimental study of free convection over finned cylinders," by R. D. Flack, *I.J.M.E.E.* 8, 2 (1980): 89.

2.B.3   "A simple laboratory apparatus for measurements of natural convection heat transfer over a flat plate in vertical, inclined, and horizontal positions," by F. F. Ling, *I.J.M.E.E.* 11, 4 (1983): 245.

2.B.4   "A transient technique for measuring the rates of heat and mass transfer to a body in a humid air flow," by I. Owen, *I.J.M.E.E.* 19, 3 (1991): 171.

2.B.5   "Convection heat transfer coefficients by the transient method," by W. M. Kays, Ref. 4, p. 118.

2.B.6   "Natural convection heat transfer coefficients," by W. M. Kays, Ref. 4, p. 121.

2.B.7   "Unsteady state convection heat transfer," by W. H. Weaver, Ref. 4, p. 176.

2.B.8   "Measurement of axial temperature distribution in a heated tube," by W. M. Rohsenow, Ref. 4, p. 178.

2.B.9   "Natural circulation loop," by S. W. Gouse, Ref. 4, p. 181.

2.B.10   "Cooling of a mercury-in-glass thermometer," by W. Rohsenow, Ref. 4, p. 230.

2.B.11   "Thermal flux meter, natural convection," by E. A. Brun, Ref. 2, p. 113.

2.B.12   "Unsteady state technique for the measurement of heat transfer coefficients," by E. A. Brun, Ref. 2, p. 114.

2.B.13   "Measurement of local heat transfer on a cylinder," by E. A. Brun, Ref. 2, p. 119.

2.B.14   "Measuring the heat produced by a single candle," by W. S. Wagner, *Phys. T.* 28 (1990): 420.

2.B.15   "How to observe convection currents in liquids," by C. L. Strong, *Sci. Am.* 216, 1 (1967): 124.

2.C   Boiling and Condensation

2.C.1   "An experimental rig for the demonstration of pool boiling," by T. A. Cowell and M. R. Heikal, *I.J.M.E.E.* 14, 2 (1986): 79.

2.D Radiation

2.D.15   "Radiometer demonstration," by J. T. Anderson, Ref. 4, p. 170.
2.D.16   "Construction of a radiometer," by P. E. Mohn, Ref. 4, p. 172.
2.D.17   "Cooling of hot water in an insulated and uninsulated can," by W. Rohsenow, Ref. 4, p. 229.
2.D.18   "Unsteady heat transfer by radiation and convection: the Rumford experiment," by E. A. Brun, Ref. 2, p. 117.
2.D.19   "Determination of emissivity," by E. A. Brun, Ref. 2, p. 118.
2.D.20   "Radiation heat transfer," by E. A. Brun, Ref. 2, p. 121.
2.D.21   "How much energy does a star radiate," by A. Harpaz, *Phys. T.* 28 (1990): 526.

2.E   Heat exchangers

2.E.1   "Regenerative heat exchangers," by S. Atallah and K. Astill, Ref. 4, p. 196.
2.E.2   "Performance of heat exchangers," by E. A. Brun, Ref. 2, p. 113.
2.E.3   "The heat pipe," by G. Y. Eastman, *Sci. Am.* 218, 5 (1968): 38.

2.F   Miscellaneous

2.F.1   "A heat transfer paradox," by A. P. Hatton and D. C. Jackson, *I.J.M.E.E.* 9, 2 (1981): 95.
2.F.2   "Heat in undergraduate education, or isn't it time we abandoned the theory of calorie," by W. F. Harris, *I.J.M.E.E.* 9, 4 (1981): 317.
2.F.3   "A computer-aided heat transfer experiment for undergraduate education," by J. Gryzagordis and K. F. Bennett, *I.J.M.E.E.* 16, 3 (1988): 189.
2.F.4   "Unsteady state heat transfer from a steam coil to water," by R. J. Aird and P. Rice, *I.J.M.E.E.* 18, 1 (1990): 37.
2.F.5   "An experimental vapour compression demonstration unit for studies on refrigeration and air conditioning," by J. R. Ghildella et al., *I.J.M.E.E.* 19, 3 (1991): 229.
2.F.6   "A project-based heat transfer course," by D. C. Anderson, *I.J.M.E.E.* 20, 2 (1992): 137.
2.F.7   "Brrr. The origin of the wind chill factor," by H. R. Crane, *Phys. T.* 27 (1989): 59.
2.F.8   "The mechanical equivalent of heat," by R. D. Edge, *Phys. T.* 25 (1987): 456.
2.F.9   "A new apparatus for measuring the mechanical equivalent of heat," by A. Saitoh, *Phys. T.* 25 (1987): 97.
2.F.10 "MHD power generation," by A. Kantrowitz and R. J. Rosa, *Phys. T.* 13 (1975): 455.
2.F.11 "Measurement of local heat transfer coefficients," by H. Cordier, Ref. 4, p. 114.

2.F.12 "Short-time response of surface temperature and heat flux measurements," by A. J. Shine, Ref. 4, p. 115.

2.F.13 "Two phase flow visualization," by Y. Y. Hsu, Ref. 4, p. 184.

2.F.14 "Evaporation of water drops from a Teflon surface," by E. Baer, Ref. 4, p. 219.

2.F.15 "Radiation and convection drying of textiles," by R. H. Wilhem, Ref. 4, p. 225.

2.F.16 "Simple mass transfer experiments," by Chemical Engr. Dept., Oregon State College, Ref. 4, p. 233.

2.F.17 "Diffusion of a vapour through stagnant air," by W. J. Heidegger, Ref. 4, p. 234.

2.F.18 "Two-penny experiments in chemical engineering," by R. Lemlich, Ref. 4, p. 235.

2.F.19 "Evaporation from perforated plates," by E. A. Brun, Ref. 2, p. 122.

2.F.20 "Velocity and temperature distribution in turbulence flow," by E. A. Brun, Ref. 2, p. 119.

2.F.21 "Heating the hard way," by S. Brusca, *Phys. T.* 28 (1990): 240.

2.F.22 "A simple apparatus for demonstration of gaseous diffusion," by R. A. Key and B. D. De Paola, *Phys. T.* 29 (1991): 522.

2.F.23 "Heat of vaporization of nitrogen," by A. W. Burgstahler and P. Hamlet, *Phys. T.* 28 (1990): 544.

2.F.24 "Diffusion experiments," by C. L. Strong, *Sci. Am.* 206, 5 (1962): 171.

2.F.25 "Conversion of sound into heat," by C. L. Strong, *Sci. Am.* 219, 2 (1968): 112.

2.F.26 "Thermal analysis technique," by C. L. Strong, *Sci. Am.* 205, 12 (1961): 170.

2.F.27 "Cooling rates of hot water," by J. Walker, *Sci. Am.* 236, 3 (1977): 246.

2.F.28 "In which heating a wire tells a lot about changes in the crystal structure of steel," by J. Walker, *Sci. Am.* 250, 5 (1984): 148.

2.F.29 "Cooking outdoors with simple equipment demonstrates aspects of thermal physics," *Sci. Am.* 253, 2 (1985): 114.

2.F.30 "Exotic patterns appear in water when it is freezing or melting," by J. Walker, *Sci. Am.* 255, 1 (1986): 114.

2.G Source texts of demonstrations and experiments

Source	Topic	# Expts.	# Demos.
Ref. 1*	Heat measurement	7	
	Conductivity	5	
	Convection	1	
	Radiation	1	
	Continuity of state	3	
Ref. 3	Conduction	19	
	Convection	7	
	Radiation	20	
Ref. 5	Conduction		10
	Heat and work		7
	Heats of transformation		18

* References are given in Appendix 1.

APPENDIX 3
Heat-transfer and thermodynamic films

Catalog number	Film title	Author(s), affiliation, and/or location	Date	Avail. from	Film type	Running time (min)
A-4	Combustion in a Small Supersonic Wind Tunnel	H. Allen, Jr. and E. A. Fletcher, NASA, Lewis	1959	C	7*	12
A-8	Electrohydrodynamic (EHD) Effects on Condensing and Evaporating Freons	P. Allen, Jr. and P. Cooper, City University, London	1987	A	6*	6
A-9	Natural Convection in a Porous Medium at High Rayleigh Numbers	M. Antos, LEPT-Ensam, URA 873 CNRS; J. P. Caltagirone and P. Fabrie, University de Bordeaux I, Talence Cedex, France	1989	A	7*	29
A-10	Boiling/Evaporative Flow Regimes in Packed Beds with Liquid–Vapor Flow of R-113	R. V. Arimilli and C. A. Moy, Department of Mechanical and Aerospace Engineering, University of Tennessee	1989	A	7*	22
B-4	Atomization Studies of Hydrazine and Nitrogen Tetroxide	M. C. Burrows, NASA, Lewis	1968	C	7*	10
B-5	Observation of Boiling by Schlieren Cinematography	M. Behar and R. Semeria, Grenoble	1964	A	5	15
B-6	Pressure Disturbances Assoc. with the Growth of Isolated Bubbles in Nucleate Boiling	H. G. Block, G. Green, G. Robinson, and F. Schmidt, Penn State U.	1973	A	1	6
B-11	Numerical Fire Simulations and Buoyant Flow Experiments	M. Baer and J. Shepherd, Sandia National Labs	1987	A	7*	17
C-4	Water Model Studies of Jet Ignition Stoker-Fired Boiler	R. W. Curtis, Babcock & Wilcox Research Center	–	A	5	14
C-9	Electrohydrodynamic Pool Boiling	H. Y. Choi, Battele Mem. Inst.	1961	A	1	4
C-12	Flow Visualization of Discrete Hole Film Cooling (C-284)	R. S. Colladay and L. M. Russell, NASA, Lewis	1976	C	7*	22
C-13	Subcooled Boiling in Normal and Zero Gravity (C-246)	T. H. Cochran and J. C. Aydelott, NASA, Lewis	1966	C	7*	11
C-14	Boiling and Dryout in Falling Thin Films	M. Cerza, National Research Council, Naval Research Laboratory, Washington, DC; V. Sernas, Rutgers University	1989	A	7*	17
E-1	Flow Visualization Studies of Free Convection Transition (on a Vertical Flat Plate)	E. R. G. Eckert et al., U. of Minn.	1958	A	1	10

ID	Title	Authors	Year			
F-2	Burnout in Flow Subcooled Boiling	M. P. Fiore and A. E. Bergles, MIT	1968	A	5	20
G-1	The Effect of Multi-g Acceleration on Nucleate Boiling Ebullition	R. W. Graham and R. C. Hendricks, NASA, Lewis	1963	A	4*	8
G-2	Pool Heating of Liquid Hydrogen in the Subcritical and Supercritical Pressure Regimes over a Range of Accelerations	R. W. Graham et al., NASA, Lewis	1963	A	7*	9
G-3	Transient Boiling in Subcooled Water and Alcohol	R. W. Graham, NASA, Lewis	1965	A	7*	10
G-4	Film Boiling and Free Convection in Carbon Dioxide near Its Critical State	R. J. Goldstein and E. Abadzic, U. of Minn	1968	A	7	16
G-5	Optical Studies of Thermal Convection	R. J. Goldstein, T. Y. Chu, and F. A. Kulaki, U. of Minn	1976	A	3	11
G-6	Heat Transfer from Heat Generating Boiling Pools	J. D. Gabor, L. Baker, Jr., and J. C. Cassulo, Argonne Nat'l Lab	1975	A	6	9
G-8	Measurement and Computation of Vorticity Structure in Turbulent Combustion (C-253)	A. F. Ghomiem and C. J. Marek, NASA, Lewis; K. Oppenheim, U. C. Berkeley	1983	C	7*	15
H-5	Two-Phase Flow in a Vertical Tube with Heat Addition	Y. Y. Hsu and R. W. Graham, NASA, Lewis	1962	C	4*	14
H-8	The Thermal Boundary Layer and the Ebullition Cycle in Nucleate Boiling	Y. Y. Hsu and R. W. Graham, NASA, Lewis	1961	C	4*	6
H-9	Film Boiling from Submerged Spheres	R. C. Hendricks and K. J. Baumeister, NASA, Lewis	1969	C	7*	22
H-11	Heat Transfer and Levitation of Fluids in Leidenfrost Film Boiling (C-267)	R. C. Hendricks, NASA, Lewis	1973	C	7*	14
K-5	Bubble Dynamics of Nucleate Boiling in Reduced Gravity	E. G. Keshock and R. Siegel	1964	A	7*	18
L-5	Stratified Flow	R. R. Long, Johns Hopkins U.	1969	B	7	26
L-7	Dominant Unstable Wavelengths in Cylindrical Interfaces	J. H. Lienhaard and P. T. Y. Wong, U. of Kentucky	1963	F	1	4

Catalog number	Film title	Author(s), affiliation, and/or location	Date	Avail. from	Film type	Running time (min)
L-9	Formation of Convecting Layers in Doubly Stratified Fluid	R. B. Lambert, U. of RI	1970	A	5	15
M-5	Flow Instabilities	E. L. Mollo-Christensen, MIT	1969	B	7	27
M-6	Holography in Heat Transfer	F. Mayinger and W. Pankin, Tech. U., Hannover, Germany	1974	A	6	18
NASA-3	Combustion Instability in a Hydrogen Oxygen Model Combustor	NASA, Lewis	1963	C	7*	14
NASA-4	Two-Phase Mercury Flow in Zero Gravity	NASA, Lewis	1963	C	7*	12
P-7	Numerical Simulations of Turbulent Natural Convection in a Square Cavity with Small Temperature Gradients	S. Paolucci, Sandia National Labs.	1987	A	6*	6
R-10	Fluid Motion in a Gravitational Field	H. Rouse and L. M. Brush, U. of Iowa	1963	H	7	24
R-15	A Study of Nucleation Sites in Boiling Using Liquid Crystals	T. Raad and J. E. Meyers, U. of California	1970	A	6	12
S-3	Flow Visualization in Combustion Systems	Shell Research Ltd.	–	A	4	12
S-13	A Visual Study of Velocity and Buoyancy Effects on Boiling Nitrogen	R. J. Simoneau and F. F. Simon, NASA, Lewis	1968	A	7*	16
S-14	Visual Evidence of an Evaporative Film underneath a Growing Bubble	R. R. Sharp and R. W. Graham, NASA, Lewis	1967	A	7*	7
S-15	Nucleate and Film Boiling in Reduced Gravity from Horizontal and Vertical Wires	R. Siegel and E. G. Keshock, NASA, Lewis	1965	A	4*	16
S-17	Boiling	R. Semaria	1961	A	5	15
S-18	Thermal Vortex Rings and Plumes	D. J. Schlien and D. W. Thompson, U. of British Columbia	1973	A	1	8
SAF-75	A Model Study of the Spread of Stratified Condenser Cooling Water at the Allen S. King Generating Plant	Northern States Power Co.	1965	J	7	10
V-1	Boiling Phenomena in Pure Liquids and Binary Mixtures	S. J. D. VanStralen, Technological U., Eidnhoven, Netherlands	1960	A	1	15

ID	Title	Author	Year	Type	No.	Min.
V-2	X-Ray Motion Pictures of Helical Mercury Flow in a Forced Flow Boiler (C-265)	A. Vary, NASA, Lewis	1970	C	7*	10
W-2	A Study of Liquid Hydrogen in Zero Gravity	L. E. Wallner and S. Nakanishi, NASA, Lewis	1963	C	7*	15
W-3	Cellular Convection in a Density Stratified Fluid Due to Lateral Heating	R. A. Wirtz, D. G. Briggs, and C. F. Chen, Rutgers, U.	1970	A	1	5
W-4	High Speed Microscopic Study of Dropwise Condensation	J. W. Westwater and J. F. Welch, U. of Ill.	1960	A	1	11
W-5	Dropwise Condensation of Ethylene Glycol	J. W. Westwater and A. C. Peterson, U. of Ill.	1965	A	1	7
W-6	Growth of Droplets in a Binary Liquid System	J. W. Westwater and R. J. Ayen, U. of Ill.	1964	A	1	10
W-7	Pool Boiling on a Glass Plate Viewed from Below	J. W. Westwater and D. B. Kirby, U. of Ill.	1963	A	1	14
W-8	Microscopic Study of Solid–Liquid Interfaces during Melting and Freezing	J. W. Westwater and L. J. Thomas, U. of Ill.	1962	A	1	16
W-9	Spontaneous Interfacial Cellular Convection Accompanying Mass Transfer	J. W. Westwater and A. Orell, U. of Ill.	1961	A	1	12
W-10	Film Boiling on a Horizontal Plate	J. W. Westwater and E. R. Hosler, U. of Ill.	1961	A	1	9
W-11	Studies of Dropwise Condensation	J. W. Westwater, U. of Ill.	1965	A	1	8
W-12	High Speed Microscopic Study of Phase Changes	J. W. Westwater and J. E. Benjamin, U. of Ill.	1960	A	1	13
W-14	Boiling Isopropanol with Trace Additives	J. W. Westwater and T. Dunskus, U. of Ill.	1960	A	1	15
W-16	Active Sites and Bubble Growth during Nucleate Boiling	J. W. Westwater and P. H. Strenge, U. of Ill.	1958	A	1	18
W-17	Film Boiling from Vertical Tubes	J. W. Westwater and Y. Y. Hsu, U. of Ill.	1957	A	1	11
W-18	A Photographic Study of Boiling	J. W. Westwater and J. G. Santangelo, U. of Ill.	1954	A	1	18
W-19	Boiling Heat Transfer from Single Fins	J. W. Westwater and K. W. Haley, U. of Ill.	1966	A	1	5
Z-2	Pool Boiling under Microgravity	M. Zell and J. Straub, Technische Universität, München	1987	A	6*	25
111649	#47 Entropy, #48 Low Temps.	ERC, U.S. Naval Acad.		M	7	60

Catalog number	Film title	Author(s), affiliation, and/or location	Date	Avail. from	Film type	Running time (min)
911650	#45 Temp./Gas Law, #46 Engines	ERC, U.S. Naval Acad.		M	7	60
811170	ABC's Auto/Diesel Engines	ERC, U.S. Naval Acad.		M	7	40
814159	Laws of Disorder. Entropy Pt. I	ERC, U.S. Naval Acad.		M	7	21
850572	Steam Power Cycle. Pt II	ERC, U.S. Naval Acad.		M	7	28
814160	Thermodynamics II	ERC, U.S. Naval Acad.		M	7	
912107	Sulzer Diesel Film	ERC, U.S. Naval Acad.		M	7	30

Index of film locations

Address	Rental fee
A. Engineering Societies Library 345 East 47th St. New York, NY 10017	Yes
B. Encyclopedia Brittanica Educational Corp. 425 North Michigan Ave. Chicago, IL 60611	Yes
C. National Aeronautics and Space Admin. Lewis Research Center Cleveland, OH 44135	No
D. National Aeronautics and Space Admin. Scientific and Technical Information Prog. Langley Research Center Hampton, VA 23365	No
E. NASA Film Loan Library Code AFEE-3 Washington, DC 20546	No
F. Mechanical Engineering Dept. University of Kentucky Lexington, KY 40506	No
G. U.S. Dept. of Commerce National Bureau of Standards Films available from: Association Sterling Films 600 Grand Ave. Ridgefield, NJ 07657	No
H. University of Iowa Media Library, Audiovisual Center Iowa City, IA 52242	Yes
I. Shell Oil Company Film Library 1433 Sadlier Cir. W. Dr. Indianapolis, IN 46239	No
J. St. Anthony Falls Hydraulic Laboratory Mississippi River at 3rd Ave. S.E. Minneapolis, MN 55414	Yes
K. Department of the Army U.S. Army Eng. Waterways Exp. Stn. P.O. Box 631 Vicksburg, MS 39180	No
L. Dept. of the Army Coastal Engineering Research Center Kingman Building Fort Belvoir, VA 22060	No
M. Educational Resource Center U.S. Naval Academy Annapolis, MD 21401-5000	No

Key to film types

Film type code	B&W or color	Silent or sound	Running speed (frames/sec)
1.	B&W	silent	16
2.	B&W	magnetic sound	16
3.	B&W	silent	24
4.	B&W	sound	24
5.	color	silent	16
6.	color	silent	24
7.	color	sound	24

Note: All films are 16 mm. Film speeds indicated are the standard running speeds although some of them contain "slow motion" sequences shot at higher speeds.

The films whose film type codes have an asterisk are available in 1/2″ VHS format and can be obtained from Engineering Society's Library.

Index

Printed in the United States
By Bookmasters